高职高专新课程体系规划教材·计算机系列

INTRODUCTORY Computer English

计算机英语基础教程

胡媛媛　编著

清华大学出版社

北　京

内 容 简 介

  本书是一本面向 21 世纪的高职高专计算机专业英语教材，内容涉及计算机基础知识、软件开发、编程语言、数据库技术、计算机网络、信息管理、动画与游戏以及其他深刻影响人们现实生活的信息技术的英语知识。本书选材广泛，以计算机和 IT 领域的最新英语时文和经典原版教材为基础，通过大量精心挑选的阅读材料，配以相应的随课注释和练习，使读者能够快速掌握计算机英语的特点和大量的术语词汇，并提高阅读和检索计算机原版文献资料的能力，特别是版式设计创意新颖，极大地方便了读者学习和查阅。书后还附有核心英语语法附录和术语、词汇索引。

  本书可供高等专科、高等职业院校计算机及 IT 相关专业的学生使用，也可供参加计算机水平考试的学生、IT 行业的工程技术人员及中等专科学校和职业高中选用。

图书在版编目（CIP）数据

计算机英语基础教程/胡媛媛编著. —北京：清华大学出版社，2011.3
（高职高专新课程体系规划教材·计算机系列）

ISBN 978-7-302-24848-4

I. ①计…   II. ①胡…   III. ①电子计算机-英语-高等学校：技术学校-教材   IV. ①H31

中国版本图书馆 CIP 数据核字（2011）第 017308 号

责任编辑：许存权 熊 健
封面设计：刘 超
版式设计：杨 洋
责任校对：张彩凤
责任印制：何 芊

出版发行：清华大学出版社   地   址：北京清华大学学研大厦 A 座
      http://www.tup.com.cn   邮   编：100084
    社 总 机：010-62770175   邮   购：010-62786544
    投稿与读者服务：010-62776969，c-service@tup.tsinghua.edu.cn
    质 量 反 馈：010-62772015，zhiliang@tup.tsinghua.edu.cn
印 装 者：北京鑫海金澳胶印有限公司
经   销：全国新华书店
开   本：185×260 印 张：15.25 字 数：352 千字
版   次：2011 年 3 月第 1 版 印 次：2011 年 3 月第 1 次印刷
印   数：1～3000
定   价：36.00 元

产品编号：041173-01

前　　言

计算机及 IT 行业是当前发展最快的领域之一，从而迫使从业人员必须快速掌握最新的前沿技术。英语作为计算机及 IT 业的行业语言，有着其他语言无可替代的功能。无论是学习最新技术，还是使用最新的计算机软硬件产品，都离不开对计算机英语的熟练掌握。正是为了适应这种要求，越来越多的高职高专院校纷纷开设了计算机英语课。有些院校不仅将该课程作为计算机专业的必修课，还将其作为一门实用的选修课推广到其他专业。

本书是在充分考虑高等职业教育的特点、学生情况、学生毕业后的就业环境、未来工作的实际要求等因素，并按照最新《大学英语教学大纲》对专业英语的要求而编写的基础教材。在满足计算机专业英语教学的同时，我们并没有过分沉溺于晦涩抽象的理论和专业术语中，而是注重实际应用与学习兴趣。全书选材广泛新颖，内容丰富，涉及了计算机基础知识、软件开发、编程语言、数据库技术、计算机网络、信息管理、动画与游戏以及其他深刻影响人们现实生活的信息技术的英语知识。

本书具有很强的实用性，课文紧扣技术潮流，把握主流经典，旨在为读者提供大量新鲜的技术术语和缩略语，为读者融入英文技术社区、阅读原文资料、提高职场竞争能力提供帮助。同时，我们在本书的体系结构编排上使之更符合计算机科学的体系构架，基本上遵循了计算机基础、软件开发、数据库技术、网络与通信、应用等 5 个层次。

本书在版式设计上创意新颖，特色的旁注生词、随文脚注难点和套色印刷，使生词更加醒目，并提高了读者对计算机英语的理解与记忆。另外，为了方便读者查阅和参考，书后还附有核心英语语法附录和术语、词汇索引。同时，我们还精心选配了插图和示意图，以增加文章内容的直观性和趣味性，进而提高读者理解原文的能力和学习兴趣。

使用说明

- 本教材根据内容分为 7 大部分，共 16 课，每一课包括两篇文章。课文为精读材料，一般为该主题的核心概述；阅读材料为泛读材料，多为该主题的某一方面或展开的讨论。两篇文章均围绕同一主题。
- 课文要求能正确理解和熟练掌握其内容；阅读材料要求能掌握中心思想，把握主要事实。
- 课文配有随课练习，包括"术语英译汉"、"知识点填空"、"理解判断"、"术语–释义搭配"、"缩略语解析"和"简答"6 种题型。各种练习均与课文内容和计算机专业紧密结合，旨在巩固和拓展学生所学内容。
- 生词均用绿色套色印刷，并在课文对应段落旁进行释义，以便阅读和记忆。相同生词原则上只在首次出现之处进行注释，但书后附有生词词汇索引以备检索。

- 计算机英语作为计算机及 IT 领域的专业英语，其特点之一就是大量专业术语的使用。本教材对其采用两种处理方式：一是在课中用斜体标出，在课后设有术语表（TERMINOLOGY）；二是对生疏的术语在文中加以脚注。但是，不论哪种情况，所有术语均收入书后术语索引，以备查阅和方便记忆。

- 对于课文中所出现的语法难点均以脚注形式加以注解，涉及语法上的通用缩写、符号则以《牛津高阶英汉双解词典（第 7 版）》为准。注释词条采用英汉双解形式，以地道的英语语言逻辑对其进行全面诠释。

- 对于教学课时比较充裕的院校，可全书使用；对教学课时较少的院校，可根据专业侧重按照自身需求进行取舍。

我们愿意给使用本书的教师和学生提供帮助（如提供练习答案、课文译文、参考试卷等）。在使用本书过程中，如存在任何问题，都可以通过电子邮件与我们交流，我们一定会给予答复。我们的 E-mail 地址：hu_yuan_yuan@yahoo.com.cn。

在本书的编写过程中，编者参考了大量教材资料，在此向这些文献资料的作者深表感谢。编者在资料的查核、术语的汉译以及文字的规范等方面都做了大量的工作，但由于计算机领域的发展日新月异，很多新的术语尚无确定的规范译法，加之水平有限，书中难免有不尽人意之处，恳请大家不吝赐教，以期共同努力，使本书成为一部"符合学生实际、切合行业实际、知识实用丰富、严谨开放创新"的优秀教材。

编　者

2010 年 12 月于石家庄

Contents

Lesson 1   What Is a Computer?

A *computer* is a device capable of performing computations and making logical decisions at speeds millions and even billions of times faster than those of human beings. For example, many of today's *personal computers* can perform hundreds of millions—even billions—of additions per second. A person operating a desk calculator might require decades to complete the same number of calculations that a powerful personal computer can perform in one second. (Points to ponder: How would you know whether the person had added the numbers correctly? How would you know whether the computer had added the numbers correctly?) Today's fastest *supercomputers* can perform hundreds of billions of additions per second— about as many calculations as hundreds of thousands of people could perform in one year! Trillion-instruction-per-second computers are already functioning in research laboratories!

Computers process *data* under the control of sets of instructions called *computer programs*. These programs guide computers through orderly sets of actions that are specified by individuals known as *computer programmers*.

A computer is composed of[1] various devices(such as the keyboard, screen[2], mouse, disks[3], memory[4], CD-ROM[5] and processing units[6]) known as *hardware*. Virtually every computer, regardless of differences in physical appearance, can be envisioned as being divided into six *logical units*, or sections:

1. *Input unit*. This "receiving" section of the computer obtains information(data and computer programs) from various *input devices*. The input unit then places this

device[di'vais]*n.*装置；设备；部件
logical['lɔdʒikəl]*adj.*符合逻辑的；按照逻辑的

ponder['pɔndə]*v.*沉思；考虑；琢磨

function['fʌŋkʃən]*v.*运转；工作
process['prəuses]*v.*数据处理
instruction[in'strʌkʃən]*n.*（计算机的）指令
orderly['ɔ:dəli]*adj.*有条理的

envision[en'vɪʒən]*v.*想象；设想

---

[1] Be composed of sth:to be made or formed from several parts, things or people;由…组成（或构成）的。
[2] Screen:also known as monitor or video display screen;显示器。
[3] Disks:also known as magnetic disks, including floppy disks and hard disks;磁盘（包括软盘和硬盘）。
[4] Memory:also known as primary storage or random access memory (RAM);主存储器，简称主存。
[5] CD-ROM (Compact Dise Read-Only Memory):光盘只读存储器，简称光盘。
[6] Processing units:also known as the central processing units (CPU);中央处理器，简称处理器。

facilitate[fə'siliteit]v.
促进;促使;使便利

audio['ɔ:djəu]n.（指方法）录音
video['vidiəu]n.（指方法）录像,录影
access['ækses]n.访问,存取（计算机文件）
temporary['tempərəri]adj.临时的;暂时的
available[ə'veiləbl]adj.可获得的;可找到的

volatile['vɔlətail]adj.不稳定的;可能急剧波动的
erase[i'reiz]v.抹去,清洗（磁带上的录音或存储器中的信息）
arithmetic[ə'riθmətik]n.算术运算,四则运算
mechanism['mekənizəm]n.方法;机制

information at the disposal the of other units[7] to facilitate the processing of the information. Today, most users enter information into computers via keyboards and mouse devices. Other input devices include microphones(for speaking to the computer), scanners(for scanning images) and digital cameras (for taking photographs and making videos).

2. *Output unit*. This "shipping" section of the computer takes information that the computer has processed and places it on various *output devices*, making the information available for use outside the computer. Computers can output information in various ways, including[8] displaying the output on screens, playing it on audio/video devices, printing it on paper or using the output to control other devices.

3. *Memory unit*. This is the rapid-access, relatively low-capacity "warehouse" section of the computer, which facilitates the temporary storage of data. The memory unit retains information that has been entered through[9] the input unit, enabling that information to be immediately available for processing. In addition[10], the unit retains processed information until that information can be transmitted to output devices. Often, the memory unit is called either *memory* or *primary memory—random access memory(RAM)*[11] is an example of primary memory. Primary memory is usually volatile, which means that it is erased when the machine is powered off.

4. *Arithmetic and logic unit(ALU)*. The ALU is the "manufacturing" section of the computer. It is responsible for the performance[12] of calculations such as addition, subtraction, multiplication and division. It also contains decision mechanisms, allowing the computer to perform such tasks as[13] determining whether two items stored in memory are equal.

5. *Central processing unit(CPU)*. The CPU serves as the "administrative" section of the computer. This is the computer's

---

[7] At the disposal of sb/sth:available for use as sb/sth prefers;任某人/某物处理，由某人/某物自行支配。
[8] Including(prep.):having sth as part of a group or set;包括…在内。
[9] Through(prep.):by means of;凭借，以…。
[10] In addition:used when you want to mention another person or thing after sth else;此外。
[11] Tip:主存储器通常分为只读存储器（Read Only Memory, ROM）、随机存储器和高速缓冲存储器（Cache）三大类。
[12] Be responsible for sth/doing sth:having the job or duty of doing sth;对…负责，有责任。
[13] Such...as...:of a kind that; like;诸如…之类。

coordinator, responsible for supervising the operation of the other sections. The CPU alerts the input unit when information should be read into the memory unit, instructs the ALU about when to use information from the memory unit in calculations and tells the output unit when to send information from the memory unit to certain output devices.

6. *Secondary storage unit* [14]. This unit is the long-term, high-capacity "warehousing" section of the computer. Secondary storage devices, such as hard drives and disks, normally hold programs or data that other units are not using; the computer then can retrieve this information when it is needed—hours, days, months or even years later. Information in secondary storage takes much longer to access than does information in primary memory. However, the price per unit of secondary storage is much less than the price per unit of

primary memory. Secondary storage is usually nonvolatileit retains information even when the computer is off.

The programs that run on a computer are referred to[15] as *software*. Software is of two major kinds:system software and application software. You can think of application software as the kind you use. Think of system software as the kind the computer uses.

System software enables the application software to interact with the computer hardware. System software is "background" software that helps[16] the computer manage its own internal resources. The most important system software program is the *operating system(OS)*, which interacts with the application software and the computer.

Application software might be described as "end-user" software. These programs are designed to address general-purpose and special-purpose applications.

Hardware costs have been declining dramatically in recent years, to the point[17] that personal computers have become a commodity. Software-development costs, however, have been

coordinator[kəu'ɔ:dɪneɪtə]*n.*协调器;协调者
supervise['sju:pəvaiz]*v.*监督;管理

retrieve[ri'tri:v]*v.*检索数据

interact[ˌintər'ækt]*v.*相互作用

address[ə'dres]*v.*处理;应对;设法解决
costs[kɔsts]*n.*成本
dramatically[drə'mætikli]*adv.*（变化、事情等）突然地;巨大地

---

[14] Secondary storage unit:also known as mass storage unit;辅助存储器。
[15] Refer to sb/sth (as sth):to mention or speak about sb/sth;提到，谈及。
[16] Help—[VN inf]。注：[动词模式代码]
[17] To the point:expressed in a simple,clear way without any extra information or feelings;简明恰当，简洁中肯。

refinement[ri'fainmənt]
n.精炼;提炼

rising steadily, as programmers develop ever more powerful and complex applications without being able to improve significantly the technology of software development. Good software-development methods can reduce software-development costs—top-down stepwise refinement[18], functionalization[19] and object—oriented programming(OOP)[20]. Object-oriented programming is widely believed to be the significant breakthrough that can greatly enhance programmer productivity.

TERMINOLOGY

application software	data
arithmetic and logic unit	disk
central processing unit	hard drive
computer	hardware
computer program	input device
computer programmer	input unit
logical unit	primary memory
memory	Random Access Memory(RAM)
memory unit	secondary storage unit
operating system	software
output device	supercomputer
output unit	system software
personal computer	

EXERCISES

1.1 Translate each of the following key terms:

    a)data
    b)application software
    c)operating system

---

[18] Top-down stepwise refinement:自顶向下逐步求精法。
[19] Functionalization:函数式软件开发法。
[20] Object-oriented programming:面向对象编程法（简称 OOP）。

d)hardware

e)supercomputer

f)central processing unit

g)computer program

h)random access memory

1.2 Fill in the blanks in each of the following statements:

a)_____ consists of the physical equipment in a computer system.

b)Computers process data under the control of sets of instructions called computer

_____.

c)CPU is an acronym for _____.

d)_____ translate the processed information from the computer into a form that humans can understand.

e)Memory, also known as _____ or_____.

f)The most common input devices are the _____ and the _____.

g)_____ is another name for programs.

h)The most important system software programs is the _____.

i)_____ holds data and program instructions for processing data.

j)Software is of two major kinds: _____ and _____.

k)Application software are designed to address _____ and _____ applications.

1.3 State whether each of the following is *true* or *false*. If *false*, explain why.

a)Good software-development methods can reduce software-development costs.

b)Hard drives is referred to as temporary storage because it is erased when the computer is powered off.

c)Programs are the instructions that tell the computer how to process data into the form you want.

d)Personal computers have become a commodity.

e)ALU is responsible for the performance of calculations such as addition, subtraction, multiplication and division.

f)Application software might be described as "end-user" software.

g)The most important kinds of secondary media are hard disks and memory.

h)The CPU serves as the "manufacturing" section of the computer.

i)The six logical sections of computer hardware:input unit, output unit, memory unit, ALU, CPU and secondary storage devices.

1.4 Categorize each of the following items as either hardware or software:

a)CPU.

b)Memory.

c)Input unit.

d)A word-processor program.

e)A Java program.

f)Windows XP.

g)Browser.

1.5 Expand each of the following acronyms:

a)ALU.

b)CPU.

c)RAM.

d)OS.

e)OOP.

Reading Material（阅读材料）

The Origins of Computing Machines

Today's computers have an extensive genealogy. One of the earlier computing devices was the *abacus*. Its history has been traced as far back as the ancient Greek and Roman civilizations. The machine is quite simple, consisting of beads strung on rods that are in turn mounted in a rectangular frame(Figure 1.1). As the beads are moved back and forth on the rods, their positions represent stored values. It is in the positions of the beads that this "computer" represents and stores data. For control of an algorithm's execution, the machine relies on the human operator. Thus the abacus alone is merely a data storage system; it must be combined with a human to create a complete computational machine.

In more recent years, the design of computing machines was based on the technology of gears. Among the inventors were Blaise Pascal(1623-1662) of France, Gottfried Wilhelm Leibniz(1646-1716) of Germany, and Charles Babbage(1792-1871) of England. These machines represented data through *gear positioning*, with data being input mechanically by establishing initial gear positions. Output from Pascal's and Leibniz's machines was achieved by observing the final gear positions. Babbage, on the other hand, envisioned machines that would print results of computations on paper so that the possibility of transcription errors would be eliminated.

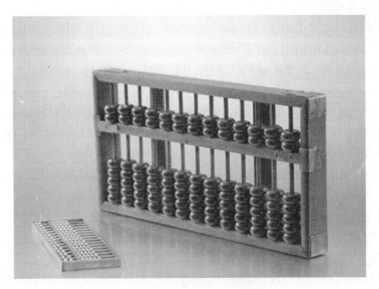

Figure 1.1   An abacus

As for the ability to follow an *algorithm*, we can see a progression of flexibility in these machines. Pascal's machine was built to perform only *addition*. Consequently, the appropriate sequence of steps was embedded into the structure of the machine itself. In a similar manner, Leibniz's machine had its algorithms firmly embedded in its architecture, although it offered a variety of *arithmetic operations* from which the operator could select. Babbage's *Difference Engine*(of which only a demonstration model was constructed) could be modified to perform a variety of calculations, but his *Analytical Engine* was designed to read instructions in the form of *holes* in *paper cards*. Thus Babbage's Analytical Engine was programmable. In fact, Augusta Ada Byron(Ada Lovelace), who published a paper in which she demonstrated how Babbage's Analytical Engine could be programmed to perform various computations, is often identified today as the world's first programmer.

The idea of communicating an algorithm via holes in paper was not originated by Babbage. He got the idea from Joseph Jacquard(1752-1834), who, in 1801, had developed a *weaving loom* in which the steps to be performed during the weaving process were determined by patterns of holes in paper cards. In this manner, the algorithm followed by the loom could be changed easily to produce different woven designs. Another beneficiary of Jacquard's idea was Herman Hollerith(1860-1929), who applied the concept of representing information as holes in paper cards to speed up the tabulation process in the 1890 U.S. census. (It was this work by Hollerith that led to the creation of IBM.) Such cards ultimately came to be known as punched cards and survived as a popular means of communicating with computers well into the 1970s. Indeed, the technique lives on today, as witnessed by the voting issues raised in the 2000 U.S. presidential election.

The technology of the time was unable to produce the complex *gear-driven machines* of Pascal, Leibniz, and Babbage in a financially feasible manner. But with the advances in electronics in the early 1900s, this barrier was overcome. Examples of this progress include the *electromechanical machine* of George Stibitz, completed in 1940 at Bell Laboratories, and the Mark I, completed in 1944 at Harvard University by Howard Aiken and a group of IBM engineers. These machines made heavy use of electronically controlled mechanical relays. In this sense, they were obsolete almost as soon as they were built, because other researchers were applying the technology of vacuum tubes to construct totally electronic computers. The first of these machines was apparently the Atanasoff-Berry machine, constructed during the period from 1937 to 1941 at Iowa State College(now Iowa State University) by John Atanasoff and his assistant, Clifford Berry. Another was a machine called Colossus, built under the direction of Tommy Flowers in England to decode German messages during the latter part of World War II. Other, more flexible machines, such as the ENIAC(electronic numerical integrator and calculator) developed by John Mauchly and J. Presper Eckert at the Moore School of Electrical Engineering, University of Pennsylvania, soon followed.

From that point on, the history of computing machines is been closely linked to advancing technology, including the invention of *transistors* and the subsequent development of *integrated circuits*, the establishment of *communication satellites*, and advances in *optic technology*. Today, small hand-held computers have more computing power than the room-size machines of the 1940s and can exchange information quickly via *global communication systems*.

A major step toward popularizing computing was the development of desktop computers. The origins of these machines can be traced to the computer hobbyists who began to experiment with homemade computers shortly after the development of the large research machines of the 1940s. It was within this "underground" of hobby activity that Steve Jobs and Stephen Wozniak built a commercially viable home computer and, in 1976, established Apple Computer, Inc., to manufacture and market their products. Other companies that marketed similar products were Commodore, Heathkit, and Radio Shack. Although these products were popular among computer hobbyists, they were not widely accepted by the business community, which continued to look to the well-established IBM for the majority of its computing needs.

In 1981, IBM introduced its first desktop computer, called the *personal computer*, or *PC*, whose underlying software was developed by a struggling young company known as Microsoft. The PC was an instant success and legitimized the desktop computer as an established commodity in the minds of the business community. Today, the term PC is widely used to refer to all those machines(from various manufacturers) whose design has evolved from IBM's initial desktop computer, most of which continue to be marketed with software from Microsoft. At times, however, the term PC is used interchangeably with the generic terms desktop or laptop.

The miniaturization of computers and their expanding capabilities have brought computer technology to the forefront of today's society. Computer technology is so prevalent now that familiarity with it is fundamental to being a member of modern society. Home computers are becoming integrated with entertainment and communication systems. *Cellular telephones* and *digital cameras* are now combined with computer technology in single hand-held units called *personal digital assistants(PDAs)*.

On a broader scale computing technology has altered the ability of governments to control their citizens, has had enormous impact on global economics, has led to startling advances in scientific research, and has repeatedly challenged society's statusquo. One can hardly imagine what the future will bring.

Note

Lesson 2   Representing Information as Bit Patterns

Inside today's computers *information* is encoded as patterns of 0s and 1s. These digits are called *bits*[1]. Although you may be inclined to associate bits with numeric values, they are really only symbols whose meaning depends on the application at hand. Thus, sometimes patterns of bits are used to represent numeric values; sometimes they represent other symbols such as characters in an alphabet and punctuation marks [2]; sometimes they represent images; and sometimes they represent sounds.

Now, we consider popular methods for encoding *text*, *numerical data*, *images*, and *sound*. Each of these systems has repercussions that are often visible to a typical computer user.

Representing Text

Information in the form of text is normally represented by means of a code in which each of the different symbols in the text(such as the letters of the alphabet and punctuation marks) is assigned a unique bit pattern. The text is then represented as a long string of bits in which the successive patterns represent the successive symbols in the original text.

In the 1940s and 1950s, many such codes were designed and used in connection with different pieces of equipment, producing a corresponding proliferation of communication problems. To alleviate this situation, the *American National Standards Institute*[3] adopted the *American Standard Code for Information Interchange*[4]. This code uses *bit patterns* of length seven to represent the *upper- and lowercase letters* of the English alphabet, punctuation symbols, the *digits* 0 through 9, and certain *control information* such as *line feeds*, *carriage returns*, and *tabs*[5]. Today, ASCII is often extended to an eight-

encode[enˈkəud]*v.* 把…编码

numerical/numeric [njuːˈmerikəl]*adj.*数字的;用数字表示的

repercussion[ˈriːpəˈkʌʃən]*n.*（间接的）影响;反响;恶果

successive[səkˈsesiv] *adj.*连续的;连接的;相继的

proliferation[prəuˌlifəˈreiʃən]*n.*激增;涌现
alleviate[əˈliːvieit]*n.*减轻;缓解

---

[1] Bits:short for binary digits;位，比特（计算机的最小信息单位）。
[2] Punctuation marks:signs or marks used in writing to divide sentences and phrases;标点符号。
[3] The American National Standards Institute:ANSI, pronounced "AN-see";美国国家标准化协会。
[4] The American Standard Code for Information Interchange:ASCII, pronounced "AS-kee";美国信息交换标准码。
[5] Line feed:换行符; carriage return:回车符; tab:制表符。

bit-per-symbol format by adding a 0 at the most significant end of each of the seven-bit patterns. This technique not only produces a code in which each pattern fits conveniently into a typical byte-size memory cell but also provides 128 additional bit patterns(those obtained by assigning the extra bit the value 1) that can represent symbols excluded in the original ASCII.[6]

Although ASCII has been the dominant code for many years, other more extensive codes, capable of representing *documents* in a variety of languages, are now competing for popularity. One of these, *Unicode*[7], was developed through the cooperation of several of the leading manufacturers of hardware and software and is rapidly gaining support in the computing community. This code uses a unique pattern of 16 bits to represent each symbol. As a result, Unicode consists of 65,536 different bit patterns—enough to allow[8] text written in such languages as Chinese, Japanese, and Hebrew to be represented.

A *file* consisting of a long sequence of symbols encoded using ASCII or Unicode is often called a *text file*. It is important to distinguish between simple text files that are manipulated by utility programs called *text editors*(or often *simply editors*) and the more elaborate files produced by *word processors*. Both consist of textual material. However, a text file contains only a character-by-character encoding of the text, whereas a file produced by a word processor contains numerous proprietary codes representing changes in *fonts*, *alignment* information, etc. Moreover, word processors may even use proprietary codes rather than a standard such as ASCII or Unicode for representing the text itself.

Representing Numeric Values

Storing information in terms of encoded characters is inefficient when the information being recorded is purely numeric. To see

dominant['dɔminənt] *adj.*占支配地位的;占优势的;显著的

manufacturer[ˌmænju'fæktʃərə]*n.*生产商;制造者

elaborate[i'læbəreit] *adj.*复杂的;详尽的

numerous['nju:mərəs] *adj.*很多的;许多的
proprietary[prə'praiəˌteri:]*adj.*专有的;所有的

---

[6] Tip: Unfortunately, because vendors tend to use their own interpretations for these extra patterns, data in which these patterns appear often are not easily transported from one vendor's application to another. （然而，由于厂商往往会对附加模式添加自己的定义和解释，因而造成了这些模式的数据很难从一厂商的应用平滑转换至另一厂商的应用中）。

[7] Tip:国际标准化组织（International Organization for Standardization, also known as ISO, in reference to the Greek word *isos*, meaning equal）已开发了一种能与 Unicode 码相竞争的新的代码标准。由于使用了 32 位模式，该编码系统足以表达几十亿不同的符号。

[8] Allow...to...[VN **to** inf]:to let sth happen or be done;允许，准许。

why, consider the problem of storing the value 25. If we insist[9] on storing it as encoded symbols in ASCII using one byte per symbol, we need a total of 16 bits. Moreover, the largest number we could store using 16 bits is 99. However, as we will shortly see, by using binary notation we can store any integer in the range from 0 to 65535 in these 16 bits. Thus, binary notation(or variations of it) is used extensively for encoded numeric data for computer storage.

Binary notation is a way of representing numeric values using only the digits 0 and 1 rather than the digits 0, 1, 2, 3, 4, 5, 6, 7, 8, and 9 as in the traditional decimal, or base ten, system. Due[10] to its efficiency, it is common to store numeric information in a form of binary notation rather than in encoded symbols. We say "a form of binary notation" because the straightforward binary system just described is only the basis for several numeric storage techniques used within machines. There are some of these variations of the

binary system, such as *two's complement notation*[11], for storing whole numbers because it provides a convenient method for representing negative numbers as well as positive, and *floating-point notation*[12], for representing numbers with fractional parts such as ⅔ or 4½.

Representing Images

Today's computer applications involve more than just text and numeric data. They include images, audio, and video. Popular techniques for representing images can be classified into two categories: *bit map* techniques[13] and *vector* techniques. In the case of bit map techniques, an image is represented as a collection of *dots*, each of which is called a *pixel*[14]. A *black and white image* is then encoded as a long string of bits representing the rows of pixels in the image, where each bit is either 1 or 0 depending on whether the corresponding pixel is black or white. This is the approach used by most facsimile machines.

shortly[ˈʃɔːtli]*adv* 立刻；马上

notation[nəuˈteiʃən]*n.*（数学、科学和音乐中用于表示信息的）符号，记号

fractional[ˈfrækʃənəl] *adj.*分数的；小数的

classify[ˈklæsifai]*v.* 将…分类；将…归类

dot[dɔt]*n.* 点；小圆点

facsimile=fax[fækˈsiməli:]*n.*传真机

---

[9] Insist on doing sth:to continue doing sth even though other people think it is annoying;执意继续做…。

[10] Due(adj.) to sth/sb:cause by sb/sth,because of sb/sth;由于，因为。

[11] Two's complement notation:二进制补码记数法。

[12] Floating-point notation:浮点记数法。

[13] Bit map techniques:位图技术;vector techniques:矢量技术。

[14] Piexl:short for "picture element";像素（picture element 的缩写）。

intensity[inˈtensiti]*n.*
强度

This bit map approach is generalized further for *color images*, where each pixel is represented by a combination of bits indicating the appearance of that pixel. Two approaches are common. In one, which we will call *RGB encoding*, each pixel is represented as three color components—a red component, a green component, and a blue component—corresponding to the three primary colors of light. One byte is normally used to represent the intensity of each color component. In turn, three bytes of storage are required to represent a single pixel in the original image.

A popular alternative to simple RGB encoding is to use a "brightness" component and two color components. In this case the "brightness" component, which is called the *pixel's* intensity, is essentially the sum of the red, green, and blue components. (Actually, it is the amount of white light in the pixel, but these details need not concern us here.) The other two components, called the *blue chrominance* and the *red chrominance*, are determined by computing the difference between the pixel's luminance and the amount of blue or red light, respectively, in the pixel. Together these three components contain the information required to reproduce the pixel.

chrominance[ˈkrəum inəns]*n.*色度（任一颜色与亮度相同的指定参考色之间的色差）

One disadvantage of bit map techniques is that an image cannot be rescaled easily to any arbitrary size. Essentially, the only way to enlarge the image is to make the pixels bigger, which leads to a grainy appearance[15]. Vector techniques provide a means of overcoming this scaling problem. Using this approach, an image is represented as a collection of *lines* and *curves*. Such a description leaves[16] the details of how the lines and curves are drawn to the device that ultimately produces the image rather than insisting that the device reproduce a particular pixel pattern.

rescale[riːˈskeil]*v.*重新调节;重新缩放
arbitrary[ˈɑːbitrəri]*adj.*粒任意的;随心所欲的
grainy[ˈgreini]*adj.*粒状的;有颗粒的

The various fonts available via today's word processing systems are usually encoded using vector techniques in order to provide flexibility in character size, resulting in scalable fonts. Vector representation techniques are also popular in *computer-*

---

[15] Tip:即"数字变焦"技术，普遍应用于数码相机；与此相对是"光学变焦"，通过调整相机镜头得到缩放效果。
[16] Leave...to...[VN +prep.]:to allow sb/sth to take care of sth;把···交托，委托。

aided design(CAD) systems in which drawings of *three-dimensional objects* are displayed and manipulated on computer screens.

Representing Sound

The most generic method of encoding *audio* information for computer storage and manipulation is to sample the amplitude of the *sound wave* at regular intervals and record the series of values obtained. For instance, the series 0, 1.5, 2.0, 1.5, 2.0, 3.0, 4.0, 3.0, 0 would represent a sound wave that rises in *amplitude*, falls briefly, rises to a higher level, and then drops back to 0. This technique, using a *sample rate* of 8000 samples per second, has been used for years in long-distance voice telephone communication. The voice at one end[17] of the communication is encoded as numeric values representing the amplitude of the voice every eight-thousandth of a second. These numeric values are then transmitted over[18] the communication line to the receiving end, where they are used to reproduce the sound of the voice.

Although 8000 samples per second may seem to be a rapid rate, it is not sufficient for high-fidelity music recordings. To obtain the quality sound reproduction obtained by today's musical CDs, a sample rate of 44,100 samples per second is used. The data obtained from each sample are represented in 16 bits(32 bits for stereo recordings). Consequently, each second of music recorded in stereo requires more than a million bits.

An alternative encoding system known as *Musical Instrument Digital Interface*[19] is widely used in the music synthesizers found in electronic keyboards, for video game sound, and for sound effects accompanying websites. By encoding directions for producing music on a synthesizer rather than encoding the sound itself, MIDI avoids the large storage requirements of the sampling technique. More precisely, MIDI encodes what instrument is to play which note for what duration of time, which

sample['saːmpl]*v.* 取样;采样

amplitude['æmpli,ruːd] *n.*（声音、无线电波等的）振幅

fidelity[fi'deliti]*n.* 高精度;高准确性

stereo['stiəriəu]*n.* 立体声

synthesizer['sinθisaizə]*n.*语音合成器

---

[17] End[usually sing.]:either of two places connected by a telephone line,etc.;端点，终端。
[18] Over:using sth;by means of sth;利用，通过。
[19] Musical Instrument Digital Interface:MIDI, pronounced "MID-ee";乐器数字化接口。

means that a clarinet playing the note D for two seconds can be encoding in three bytes rather than more than two million bits when sampled at a rate of 44,100 samples per second.

In short, MIDI can be thought of as a way of encoding the sheet music read by a performer rather than the performance itself, and in turn, a MIDI "recording" can sound significantly different when performed on different synthesizers.

TERMINOLOGY

alignment	information
American National Standards Institute	line
American Standard Code for Information	line feed
amplitude	lowercase letter
audio	Musical Instrument Digital Interface
binary notation	numerical data
bit	pixel
bit map	pixel's luminance
bit pattern	RGB encoding
black and white image	sample rate
blue chrominance	sound
carriage return	sound wave
color image	tab
control information	three-dimensional object
computer-aided design	text
curve	text editor/simply editor
digit	text file
dot	two's complement notation
file	Unicode
floating-point notation	uppercase letter
font	vector
image	word processor

EXERCISES

2.1 Translate each of the following key terms:

a)pixel
b)lowercase letter
c)image
d)file
e)bit map
f)audio
g)curve
h)sample rate

2.2 Fill in the blanks in each of the following statements:

a)Computer information is encoded as patterns of 0s and 1s, each of which is called

_____.

b)ASCII stands for _____.

c)Popular techniques for representing images can be classified into two
categories:_____ techniques and _____ techniques.

d)MIDI is an acronym for _____.

e)In order to represent numbers, letters, and special characters, bits are combined
into groups of eight bits called _____.

f)A bitmap image is represented as a collection of dots, each of which is called a

_____.

g)_____ is another name for text editors.

h)The most common method of encoding audio is to sample the _____ of
the sound wave at regular intervals and record the series of values.

i)Unicode is a _____-bit code designed to support international languages like
Chinese and Japanese.

j)A vector image is represented as a collection of _____ and _____.

2.3 State whether each of the following is *true* or *false*. If *false*, explain why.

a)The American National Standards Institute adopted Unicode to alleviate the
proliferation of communication problems.

b)A file consisting of a long sequence of symbols encoded using ASCII or Unicode is
often called a text file.

c)The floating-point notation is used for storing numbers for fractional parts.

d)ASCII is the most widely used decimal code for computers.

e)Storing information in terms of encoded characters is efficient whether the information being recorded is purely numeric or not.

f)A MIDI "recording" can sound significantly different when performed on different synthesizers.

g)One disadvantage of bit map techniques is that an image cannot be rescaled easily to any arbitrary size.

h)When you press a key on the keyboard, a character automatically converted into a series of bits.

i)To obtain the quality sound reproduction obtained by today's musical CDs, a sample rate of 8,000 samples per second is used.

2.4 Match each numbered item with the most closely related lettered item:

a)binary system	1. a binary system for representing whole integer numbers.
b)vector techniques	2. consists of rows of pixels, white or black.
c)two's complement notation	3. a numbering system that consists of only two digits— 0 and 1.
d)black and white image	4. popular techniques for representing images.
e)control information	5. such as line feeds, carriage returns, and tabs.

2.5 Expand each of the following acronyms:

    a)ASCII.
    b)CAD.
    c)MIDI.
    d)RGB.
    e)piexl.

Reading Material （阅读材料）

Generic Data Compression Techniques

For the purpose of storing or transferring data, it is often helpful(and sometimes mandatory) to reduce the size of the data involved while retaining the underlying information. The technique for accomplishing this is called *data compression*. Let us consider some generic data compression methods.

    Data compression schemes fall into two categories. Some are *lossless*, others are *lossy*. Lossless schemes are those that do not loose information in the compression process. Lossy schemes are those that may lead to the loss of information. Lossy techniques often provide more compression than losseless ones and are therefore popular in settings in which minor errors can be tolerated, as in the case of images and audio.

In cases where the data being compressed consist of long sequences of the same value, the compression technique called *run-length encoding*, which is a lossless method, is popular. It is the process of replacing sequences of identical data elements with a code indicating the element that is repeated and the number of times it occurs in the sequence. For example, less space is required to indicate that a bit pattern consists of 253 ones, followed by 118 zeros, followed by 87 ones than to actually list all 458 bits.

Another lossless data compression technique is *frequency-dependent encoding*, a system in which the length of the bit pattern used to represent a data item is inversely related to the frequency of the item's use. Such codes are examples of variable-length codes, meaning that items are represented by patterns of different lengths as opposed to codes such as Unicode, in which all symbols are represented by 16 bits. David Huffman is credited with discovering an algorithm that is commonly used for developing frequency-dependent codes, and it is common practice to refer to codes developed in this manner as *Huffman codes*. In turn, most frequency-dependent codes in use today are Huffman codes.

As an example of frequency-dependent encoding, consider the task of encoded English language text. In the English language the letters e, t, a, and i are used more frequently than the letters z, q, and x. So, when constructing a code for text in the English language, space can be saved by using short bit patterns to represent the former letters and longer bit patterns to represent the latter ones. The result would be a code in which English text would have shorter representations than would be obtained with uniform-length codes.

In some cases, the stream of data to be compressed consists of units, each of which differs only slightly from the preceding one. An example would be consecutive frames of a motion picture. In these cases, techniques using *relative encoding*, also known as *differential encoding*, are helpful. These techniques record the differences between consecutive data units rather than entire units; that is, each unit is encoded in terms of its relationship to the previous unit. Relative encoding can be implemented in either lossless or lossy form depending on whether the differences between consecutive data units are encoded precisely or approximated.

Still other popular compression systems are based on *dictionary encoding* techniques. Here the term dictionary refers to a collection of building blocks from which the message being compressed is constructed, and the message itself is encoded as a sequence of references to the dictionary. We normally think of dictionary encoding systems as lossless systems, but as we will see in our discussion of image compression, there are times when the entires in the dictionary are only approximations of the correct data elements, resulting in a lossy compression system.

Dictionary encoding can be used by word processors to compress text documents because the dictionaries already contained in these processors for the purpose of spell checking make excellent compression dictionaries. In particular, an entire word can be encoded as a single reference to this dictionary rather than as a sequence of individual characters encoded using a system such as ASCII or Unicode. A typical dictionary in a word processor contains approximately 25,000 entires, which means an individual entry can be identified by an integer in the range of 0 to 24,999. This means that a particular entry in the dictionary can be identified by a pattern of only 15 bits. In contrast, if the word being referenced consisted of six letters, its character-by-character encoding would require 42 bits using seven-bit ASCII or 96 bits using Unicode.

A variation of dictionary encoding is *adaptive dictionary encoding*(also known as dynamic dictionary encoding). In an adaptive dictionary encoding system, the dictionary is allowed to change during the encoding process. A popular example is *Lemple-Ziv-Welsh(LZW) encoding*(named after its creators, Abraham Lemple, Jacob Ziv, and Terry Welsh). To encode a message using LZW, one starts with a dictionary containing the basic building blocks from which the message is constructed, but as larger units are found in the message, they are added to the dictionary—meaning that future occurrences of those units can be encoded as single, rather than multiple, dictionary references. For example, when encoding English text, one could start with a dictionary containing individual characters, digits, and punctuation marks. But as words in the message are identified, they could be added to the dictionary. Thus, the dictionary would grow as the message is encoded, and as the dictionary grows, more words(or recurring patterns of words) in the message could be encoded as single references to the dictionary.

The result would be a message encoded in terms of a rather large dictionary that is unique to that particular message. But this large dictionary would not have to be present to decode the message. Only the original small dictionary would be needed. Indeed, the decoding process could begin with the same small dictionary with which the encoding process started. Then, as the decoding process continues, it would encounter the same units found during the encoding process, and thus be able to add them to the dictionary for future reference just as in the encoding process.

To clarify, consider applying LZW encoding to the message

xyx xyx xyx xyx

starting with a dictionary with three entries, the first being x, the second being y, and the third being a space. We would begin by encoding xyx as 121, meaning that the message starts with the pattern consisting of the first dictionary entry, followed by the second, followed by the first. Then the space is encoded to produce 1213. But, having reached a space, we know that the preceding string of characters forms a word, and so we add the pattern xyx to the dictionary as the fourth entry. Continuing in this manner, the entire message would be encoded as 121343434.

If we were now asked to decode this message, starting with the original three-entry dictionary, we would begin by decoding the initial string 1213 as xyx followed by a space. At this point we would recognize that the string xyx forms a word and add it to the dictionary as the fourth entry, just as we did during the encoding process. We would then continue decoding the message by recognizing that the 4 in the message refers to this new fourth entry and decode it as the word xyx, producing the pattern

     xyx xyx.

Continuing in this manner we would ultimately decode the string 121343434 as

     xyx xyx xyx xyx

which's the original message.

Note

Lesson 3　What Is an Operating System?

An operating system is an important part of almost every computer system. A *computer system* can be divided roughly into four components:the *hardware*, the *operating system*, the *applications programs*, and the *users*(Figure 3.1).

roughly['rʌfli]*adv.*粗略地;大体上;大致上

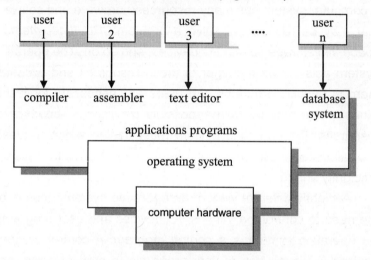

Figure 3.1　Abstract view of the components of a computer system

The hardware—the central processing unit(CPU), the memory, and the input/output(I/O) devices—provides the basic computing resources. The applications programs—such as compilers, database systems, games, and business programs—define the ways in which[1] these resources are used to solve the computing problems of[2] the users. There may be many different users(people, machines, other computers) trying[3] to solve different problems. Accordingly, there may be many different applications programs. The operating system controls and coordinates the use of[4] the hardware among the various applications programs for the various users.

accordingly[ə'kɔ:diŋli]
adv.（尤其用于句首）因此;于是
coordinate[kəu'ɔ:din eit]*v.*使…得以协调

---

[1]　Tip:Here the relative pronoun "which" refers to "the ways", and the subject of the relative clause is "these resources".
[2]　Of:belonging to sb; relating to sb;属于（某人）；关于（某人）。
[3]　同:...many different users who try to solve...
[4]　Of:used after nouns formed from verbs. The noun after "of" can be either the object or the subject of the action;用于由动词转化的名词之后，of 之后的名词可以是受动者，也可以是施动者。

An operating system is similar to[5] a government. The components of a computer system are its hardware, software, and data. The operating system provides the means for the proper use of these resources in the operation of the computer system. Like a government, the operating system performs no useful function by itself. It simply provides an environment within which other programs can do useful work.

We can view an operating system as[6] a *resource allocator.* A computer system has many resources(hardware and software) that may be required to solve a problem:*CPU time*, *memory space*, *file storage space*, *I/O devices*, and so on. The operating system acts as the manager of these resources and allocates them to specific programs and users as[7] necessary for tasks. Since[8] there may be many—possibly conflicting—requests for resources, the operating system must decide which requests are allocated resources to operate the computer system efficiently and fairly.

A slightly different view of an operating system focuses on the need to control the various I/O devices and user programs. An operating system is a *control program*. A control program controls the execution of user programs to prevent errors and improper use of the computer. It is especially concerned with the operation and control of I/O devices. The view holds that the operating system is there to manage all the pieces of a complex system. So, the job of the operating system is to provide for an orderly[9] and controlled allocation of the processors, memories, and I/O devices among the carious programs competing for them.

In general, however, there is no completely adequate definition of an operating system. Operating systems exist because they are a reasonable way to solve the problem of[10] creating a usable computing system. The fundamental goal of computer systems is to execute user programs and to make

specific[spi'sifik]*adj.*
（名词前）特定的
conflicting[kən'fliktiŋ]
*adj.*相互矛盾的;相冲突的
request[ri'kwest]*n.*
（计算机的）请求
fairly['fɛəli]*adv.*公平地,公正地

carious['kɛəriəs]*adj.*
（力量、影响等）衰弱,衰败;精疲力尽的

---

[5] Be similar to sb/sth| be similar in sth:like sb/sth but not exactly the same;相像的；类似的。
[6] View sb/sth as sth|~ sb/sth with sth:to think about sb/sth in a particular way;把…视为；以…看待。
[7] As(adv.):used to state the reason for sth;因为；由于。
[8] Since(conj.):because;as;因为；由于。
[9] Orderly(adj.):arranged or organized in a careful and logical way;有条理的；有秩序的。
[10] Of:used to say what sth is, consists of, or contains;（用于表示性质、组成或涵盖）即，由…组成。

solving user problems easier. Toward this goal, computer hardware is constructed. Since bare hardware alone is not particularly easy to use, applications programs are developed. These various programs require certain[11] common operations, such as those controlling the I/O devices. The common functions of controlling and allocating resources are then brought together into one piece of software:the operating system.

While there are hundreds of different operating systems, there are only three basic categories:embedded, network, or stand-alone.

Embedded operating systems are used for handheld computers and smaller devices like PDAs. These operating systems are called embedded because they are completely stored within the device in its ROM memory. Popular embedded operating systems include Windows CE and Palm OS.

handheld['hænd,held]
*adj.*手持的

Network operating systems(NOS) are used to control and coordinate computers that are linked together. The operating system typically is located on one of the connected computers' hard disks. Called the *network server*, this computer coordinates all communication between the other computers. Popular network operating systems include NetWare, Windows NT Server, Windows XP Server, and UNIX.

Stand-alone operating systems, also called *desktop operating systems*, control a single desktop or notebook computer. These operating systems are located on the computer's hard disk. Often desktop computers and notebooks are part[12] of a network. In these cases, the desktop operating system works with the network's NOS to share and coordinate resources. In these situations, the desktop operating system is referred to[13] as the *client operating system*. Popular desktop operating systems include Windows, Mac OS, and some versions of UNIX.

---

[11] Certain(adj.):used to mention a particular thing, person or group without giving any more details about it or them;（不提及细节时用）某…。

[12] Part[U] of sth:some but not all of a thing;部分。

[13] Refer to sb/sth as sth:to mention or speak about sb/sth;提到；谈及；说起。

universally[ˌjuːniˈvəː
səlɪ]adv.全体地;一致地
vendor[ˈvendə]n.销
售商
ship[ʃip]v.上市;把…推
向市场
lack[læk]v.缺少

There is also no universally accepted definition of what is part of the operating system and what is not. A simple viewpoint is that everything a vendor ships when you order "the operating system" should be considered. The memory requirements and features included, however, vary greatly across systems. Some take up less than 1 megabyte[14] of space and lack even a full-screen editor, whereas[15] others require hundreds of megabytes of space and include *spell checkers* and entire "window systems". A more common definition is that the operating system is the one program running at all times on the computer (usually called the *kernel*), with all else being applications programs. This last definition is more common and is the one we generally follow.

It[16] is easier to define operating systems by what they do than by what they are. The primary goal of an operating system is convenience for the user. Operating systems exist because they are supposed to make it easier to compute with them than without them. This view is particularly clear when you look at operating systems for small personal computers.

A secondary goal is efficient operation of the computer system. This goal is particularly important for large, shared *multiuser systems*. These systems are typically expensive, so it is desired to make them as efficient as possible. These two goals—convenience and efficiency—are sometimes contradictory. In the past, efficiency considerations were often more important than convenience. Thus, much of operating system theory concentrates on optimal use of computing resources.

contradictory[ˌkɔntrə
ˈdiktəriː]adj.相互矛盾
的;对立的
concentrate[ˈkɔnsənt
reit]v.集中（注意力）;
聚精会神
optimal[ˈɔptəməl]adj.
最佳的;最优的

To see what operating systems are and what operating systems do, let us consider how they have developed over[17] the past 35 years. By tracing that evolution, we can identify the common elements of operating systems, and see how and why these systems have developed as they have.

---

[14] Megabyte:1 megabyte is a million bytes;兆字节。

[15] Whereas:used to compare or contrast two facts; （用以比较或对比两个事实）然而，但是，尽管。

[16] It:used in the position of the subject or object of a verb when the real subject or object is at the end of sentence;用作形式主语或形式宾语，而真正的主语或宾语在句末。

[17] Over:more than a particular time, amount, cost, etc.;多于（某时间、数量、花费等）。

Operating systems and *computer architecture* have had a great deal[18] of influence on each other. To facilitate the use of the hardware, operating systems were developed. As operating systems were designed and used, it because obvious that changes in the design of the hardware could simplify them. In this short historical review, notice how operating-system problems led to the introduction of new hardware features.

facilitate[fə'siliteit]*v.*
（正式的）促进;使便利

TERMINOLOGY

applications program	kernel
client operating system	memory space
computer architecture	multiuser system
computer system	network operating system(NOS)
control program	network server
CPU time	operating system
desktop operating system	resource allocator
embedded operating system	spell checker
file storage space	stand-alone operating system
hardware	user
I/O device	

EXERCISES

3.1 Translate each of the following key terms:

 a)computer system

 b)desktop operating system

 c)operating system

 d)control program

 e)multiuser system

 f)resource allocator

---

[18] A great/good deal:much; a lot;大量；很多。

g)embedded operating system

h)kernel

3.2 Fill in the blanks in each of the following statements:

a)A computer system can be divided into four components:the hardware, the _____, the applications programs, and the users.

b)Operating systems are classified into three basic categories:embedded, network, or _____.

c)_____ are programs that coordinate computer resources, provide an interface between users and the computer, and run applications.

d)A(n) _____ controls the execution of user programs to prevent errors and improper use of the computer.

e)A computer that coordinates all communication between other computers is called a _____.

f)Desktop operating systems are also called _____ operating systems.

g)Embedded operating systems are completely stored with _____ memory.

h)I/O is an acronym for _____.

i)The _____ operating system was originally designed to run on minicomputers in network environments.

j)In network situations, the desktop operating system is referred to as the _____.

3.3 State whether each of the following is *true* or *false*. If *false*, explain why.

a)Operating systems control and coordinate the use of the hardware among the various applications programs for the various users.

b)Every operating system performs three basic functions:managing resources, providing a user interface, and running applications.

c)Stand-alone operating systems are used for handheld computers and smaller devices.

d)A desktop operating system is an operating system located on a single stand-alone hard disk.

e)1 megabyte is referred to as a unit of information equal to one million million, or 10^{12} bytes.

f)Linux is one popular, and free, version of the UNIX operating system.

g)The Mac OS operating system is designed to run on Intel and Intel-compatible micropeocessors.

3.4 Categorize each of the following items as embedded, network, or stand-alone operating systems:

a)Windows.

b)UNIX.

c)Palm OS.

d)Windows XP Server.

e)NetWare.

f)Windows CE.

g)Mac OS.

h)Windows NT Server.

i)Linux.

3.5 Expand each of the following acronyms:

a)NOS.

b)CPU.

c)ROM.

d)OS.

e)PDA.

Reading Material （阅读材料）

Windows Vista

Windows is a modern operating system that runs on consumer and business desktop PCs and enterprise servers. The most recent desktop version is *Windows Vista*. The server version of Windows Vista is called *Windows Server 2008*.

Microsoft's development of the *Windows operating system* for PC-based computers as well as servers can be divided into three eras:*MS-DOS, MS-DOS-based Windows*, and *NT-based Windows*. Technically, each of these systems is substantially different from the others. Each of these was dominant during different decades in the history of the personal computer. Figure 3.2 shows the dates of the major Microsoft operating system releases for desktop computers(omitting the popular Microsoft Xenix version of UNIX, which Microsoft sold to the Santa Cruz Operation(SCO) in 1987). Below we will briefly sketch the Windows Vista era.

Year	MS-DOS	MS-DOS-based Windows	NT-based Windows	Notes
1981	MS-DOS 1.0			Initial release for IBM PC
1983	MS-DOS 2.0			Support for PC/XT
1984	MS-DOS 3.0			Support for PC/AT
1990		Windows 3.0		Ten million copies in 2 years
1991	MS-DOS 5.0			Added memory management
1992		Windows 3.1		Runs only on 286 and later
1993			Windows NT 3.1	
1995	MS-DOS 7.0	Windows 95		MS-DOS embedded in Win 95
1996			Windows NT 4.0	
1998		Windows 98		
2000	MS-DOS 8.0	Windows Me	Windows 2000	Win Me was inferior to Win 98
2001			Windows XP	Replaced Windows 98
2006			Windows Vista	

Figure 3.2   Major releases in the history of Microsoft operating systems for desktop PCs

The *release* of Windows Vista culminated Microsoft's most extensive operating system project to date. The initial plans were so ambitious that a couple of years into its development Vista had to be restarted with a small scope. Plan to rely heavily on Microsoft's type-safe, garbage-collected .NET language *C#* were shelved, as were some significant features such as the *WinFS unified storage system* for searching and organizing data from many different sources. The size of the full operating system is staggering. The original NT release of 3 million lines of C/C++ that had grown to 16 million in NT 4, 30 million in 2000, and 50 million in XP, is over 70 million lines in Vista.

Much of the size is due to Microsoft's emphasis on adding many new features to its products in every release. In the main *system32* directory, there are 1600 *dynamic link libraries(DLLs)* and 400 *executables(EXEs)*, and that does not include the other directories containing the myriad of applets included with the operating system that allow users to surf the Web, play music and video, send e-mail, scan documents, organize photos, and even make movies. Because Microsoft wants customers to switch to new versions, it maintains compatibility by generally keeping all the features, APIs, *applets*(small applications), etc., from the previous version. Few things ever get deleted. The result is that Windows grows dramatically release to release. Technology has kept up, and Windows' distribution media have moved from floppy, to CD, and now with Windows Vista, DVD.

The bloat in features and applets at the top of Windows makes meaningful size comparisons with other operating systems problematic because the definition of what is or is not part of an operating system is difficult to decide. At the lower layers of operating systems, there is more correspondence because the functions performed are very similar. Even so we can see a big difference in the size of Windows. Figure 3.3 compares the Windows and Linux *kernels* for three key functional areas: *CPU scheduling*, *I/O infrastructure*, and *Virtual Memory*. The first two components are half again as large in Windows, but the Virtual memory component is an order of magnitude larger-duo to the large number of features, the virtual memory model used, and implementation techniques that trade off code size to achieve higher performance.

Kernel area	Linux	Vista
CPU Scheduler	50,000	75,000
I/O infrastructure	45,000	60,000
Virtual Memory	25,000	175,000

Figure 3.3   Comparison of lines of code for selected kernel-mode modules in Linux and Windows

Kernel mode in Windows Vista is structured in the *HAL*, the kernel and executive layers of *NTOS*, and a large number of device drivers implementing everything from device services to files systems and networking to graphics. The HAL hides certain differences in hardware from the other components. The kernel layer manages the CPUs to support *multithreading* and *synchronization*, and the executive implements most *kernel-mode services*.

The executive is based on kernel-mode objects that represent the key executive data structures, including processes, threads, memory sections, drivers, devices, and synchronization objects—to mention a few. User processes create objects by calling system services and get back handle references which can be used in subsequent system calls to the executive components. The operating system also creates objects internally. The object manager maintains a name space into which objects can be inserted for subsequent lookup.

The most important objects in Windows are *processes*, *threads*, and *sections*. Processes have *virtual address spaces* and are containers for resources. Threads are the unit of execution and are scheduled by the kernel layer using a priority algorithm in which the *highest-priority* ready thread always runs, preempting *lower-priority* threads as necessary. Sections represent memory objects, like *files*, that can be mapped into the address spaces of processes. EXE and DLL program images are represented as sections, as is shared memory.

Windows supports *demand-paged virtual memory*. The paging algorithm is based on the *working-set* concept. The system maintains several types of page lists, to optimize the use of memory. The *various page lists* are fed by trimming the working sets using complex formulas that try to reuse physical pages that have not been referenced in a long time. The *cache manager* manages virtual addresses in the kernel that can be used to map files into memory, dramatically improving I/O performance for many applications because read operations can be satisfied without accessing disk.

I/O is performed by *device driver*s, which follow the *Windows Driver Model*. Each driver starts out by initializing a driver objects that contains the address of the procedures that the system can call to manipulate devices. The actual devices are represented by device objects, which are created from the configuration description of the system or by the *plug-and-play* manger as it discovers devices when enumerating the system buses. Devices are stacked and I/O request packets are passed down the stack and serviced by the drivers for each device in the device stack. I/O is inherently asynchronous, and drivers commonly queue requests for further work and return back to their caller. *File system volumes* are implemented as devices in the I/O system.

The *NTFS file system* is based on a master file table, which has one record per file or directory. All the metadata in an NTFS file system is itself part of an NTFS file. Each file has multiple attributes, which can either be in the MFT record or nonresident(stored in blocks outside the MFT). NTFS supports Unicode, compression, journaling, and encryption among many other features.

Finally, Windows Vista has a sophisticated security system based on *access control lists* and *integrity levels*. Each process has an *authentication token* that tells the identity of the user and what special privileges the process has, if any. Each objects has a security descriptor associated with it. The security descriptor points to a discretionary access control list that contains access control entries that can allow or deny access to individuals or groups. Windows has added numerous security features in recent releases, including BitLocker for encrypting entire volumed, and address space randomization, nonexecutable stacks, and other measures to make buffer overflow attacks more difficult.

Note

Lesson 4　Overview of Systems Development

objective[əb'dʒektiv]
*n.*目标;目的
solving['sɔlviŋ]*n.*解决;
处理;求解

Whatever their scope and objectives, new *software systems* are an outgrowth of[1] a process of[2] application problem solving. *Systems development* refers to all[3] the activities that go into producing an software systems solution to an application problem or opportunity. Systems development is a structured kind of problem solving with distinct activities. These activities consist of[4] systems analysis, systems design, programming, testing, conversion, and production and maintenance.

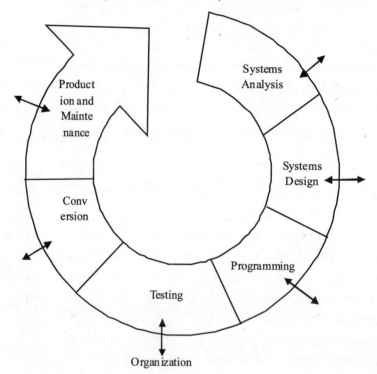

Figure 4.1　The systems development process. Each of the core systems development activities entails interaction with the organization.

---

[1] Of:used after nouns formed from verbs. The noun after "of" can be either the object or the subject of the action;用于由动词转化的名词之后，of 之后的名词可以是受动者，也可以是施动者。

[2] Of:concerning or showing sb/sth;关于，反映（某人或某事）。

[3] All(det.):(used with plural nouns. The noun may have *the* , *this*, *that*, *my*, *her*, *his*, etc. in front of it, or a number.) the whole number of;所有；全部；全体。

[4] Of:used after some verbs before mentioning sb/sth involved in the action;用于某些动词后，后接动作所涉及的人或事。

Figure 4.1 illustrates the *systems development process*. The systems development activities depicted here usually take place[5] in sequential order. But some of the activities may need to be repeated or some may be taking place simultaneously, depending on the approach[6] to system building that is being employed. Note also that each activity involves interaction with the users. Users participate in these activities, and the systems development process creates *requirement changes*.

Systems Analysis

System analysis is the analysis of the problem that[7] a organization will try to solve with a software system. It consists of defining the problem, identifying its causes, specifying the solution, and identifying the *information requirements* that must be met by a *system solution*.

The systems *analyst* creates a road map of the existing organization and systems, identifying the primary owners and users of data in the organization. These stakeholders have a direct interest in the information affected[8] by the new system. In addition to these organizational aspects, the analyst also briefly describes the existing hardware and software that serve the organization.

From this organizational analysis, the system analyst details the problems or limitations of[9] existing systems. By examining documents, work papers, and procedures; observing system operations; and interviewing key users of the systems, the analyst can identify the problem areas and objectives to be achieved by a solution. Often the solution requires[10] building a new system or improving an existing one.

In addition to suggesting a solution, systems analysis involves a *feasibility study* to determine whether that solution is feasible, or achievable, given[11] the organization's resources and constraints. Three major areas of feasibility must be

illustrate['iləstreit]v.（用示例、图画等）说明,解释

depict[di'pikt]v.描写；描述

participate[pɑ:'tisipeit] v.（正式的）参加;参与

existing[ig'zistiŋ]adj. 现存的

stakeholder['steikhə uldə]n.（某组织、工程、体系等的）参与者，参与方；有权益关系者

feasible['fi:zəbl]adj. 可行的;行得通的

---

[5] Take place:to happen, especially after previously being arranged or planned;（尤指根据安排或计划）发生，进行。

[6] Approach[C] to sth:a way of dealing with sb/sth; a way of doing or thinking about sth such as a problem or a task;（待人接物或思考问题的）方式，方法，态度。

[7] Tip:Here the relative pronoun "that" refers to "the problem", and the subject of the relative clause is "a organization".

[8] 同:...the information that is affected by the new system.

[9] Of:belonging to sth; being part of sth; relating to sth;属于（某物）；关于（某物）。

[10] Require—[V -ing]。

[11] Given(prep.):when you consider sth;考虑到；鉴于。

address[əˈdres]v. 设法
解决;处理

outweigh[autˈwei]v.大
于;超过

framework[ˈfreimwəːk]
n.（体系的）结构,机制;
架构

pose[pəuz]v.造成（问
题、威胁等）;引起

addressed:

1. *Technical feasibility*:whether the proposed solution can be implemented with the available hardware, software, and technical resources.
2. *Economic feasibility*:whether the benefits of the proposed solution outweigh the costs.
3. *Operational feasibility*:whether the proposed solution is desirable within the existing managerial and organizational framework.

Perhaps the most difficult task of the systems analyst is to define the specific information requirements that must be met by the system solution selected. This is the area where many large system efforts go wrong and the one that poses the greatest difficulty for the analyst. At the most basic level, the *information requirements* of a new system involve identifying who needs what information, where, when, and how. *Requirements analysis* carefully defines the objectives of the new or modified system and develops a detailed description of the functions that the new system must perform. Requirements must consider economic, technical, and time constraints, as well as the goals, procedures, and decision processes of the organization. Faulty requirements analysis is a leading[12] cause of *systems failure* and high *systems development costs*.

Systems Design

fulfill[fulˈfil]v.履行;实现
overall[ˈəuvərɔːl]adj.
全面的;总体的

deliver[diˈlivə]v.清楚
表述

Whereas[13] systems analysis describes what a system should do to meet information requirements, *system design* shows how the system will fulfill this objective. The design of a software system is the overall plan or model for that system. Like the blueprint of a building or house, it consists of all specifications that give[14] the system its form and structure.

The *system designer* details the system specifications that will deliver the functions identified during systems analysis. These specifications should address all the managerial, organizational, and technological components of the system solution. Table 4.1 lists the types of specifications that would be produced during

---

[12] Leading(adj.)[only before noun]:most important;最重要的。
[13] Whereas:used to compare or contrast two facts;（用以比较或对比两个事实）然而，但是，尽管。
[14] Give—[VNN]。

system design.

Table 4.1   Design Specifications

Output	Controls
Medium	Input controls(characters, limit, reasonableness)
Content	Processing controls(consistency, record counts)
Timing	Output controls(totals, samples of output)
Input	Procedural controls(passwords, special forms)
Origins	Security
Flow	Access controls
Data entry	Catastrophe plans
User interface	Audit trails
Simplicity	Documentation
Efficiency	Operations documentation
Logic	System documents
Feedback	User documentation
Errors	Conversion
Database design	Transfer files
Logical data relations	Initiate new procedures
Volume and speed requirements	Select testing method
File organization and design	Cut over to new system
Record specifications	Training
Processing	Select training techniques
Computations	Develop training modules
Program modules	Identify training facilities
Required reports	Organizational changes
Timing of outputs	Task redesign
Manual procedures	Job design
What activities	Process design
Who performs them	Office and organization structure design
When	Reporting relationships
How	
Where	

The design for an new system can be broken down into[15] logical and physical design specifications. Logical design lays out the components of the system and their relationship to[16] each other as they would appear to users. It shows what the system solution will do as opposed to[17] how it is actually

---

[15] Break sth down into sth:to separate sth into smaller parts in order to analyse it or deal with it more easily;把…分类；划分。

[16] To:used to show a relationship between one person or thing and another; （表示两人或事物之间的关系）属于，关于，对于。

[17] As opposed to:used to make a contrast between two things; （表示对比）而，相对于。

implemented physically. It describes inputs and outputs, processing functions to be performed, business procedures, data models, and controls.

Physical design is the process of translating the abstract logical model into the specific technical design for the new system. It produces the actual specifications for hardware, software, physical databases, input/output media, manual procedures, and specific controls. Physical design provides the remaining specifications that transform the abstract logical design plan into a function system of people and machines.

remaining[ri'meiniŋ]
*adj.*其余的

Programming

The process of translating design specifications into software for the computer constitutes a smaller portion of the systems development cycle than design and, perhaps, the testing activities[18]. During the programming stage, system specifications that were prepared during the design stage are translated into program code. On the basis of detailed design documents for files, transaction and report layouts, and other design details, specifications for each program in the system are prepared.

portion['pɔ:ʃən]n.部分

Testing

Exhaustive and thorough testing must be conducted to ascertain whether the system produces the right results. Testing answers the questions, "Will the system produce the desired results under known conditions?"

The amount of time needed to answer this question has been traditionally underrated in systems project planning. As much as 50 percent of the entire software development budget can be expended in testing. Test data must be carefully prepared, results reviewed, and corrections made in the system[19]. In some instances, parts of the system may have to be redesigned. Yet the risk of glossing over this step are enormous. Testing a new system can be divided into three types of activities:unit testing, system testing, and acceptance testing.

exhaustive[ig'zɔ:stiv]
*adj.*详尽的;彻底的;全面的
thorough['θʌrə]*adj.* 深入的;细致的
ascertain['æsə'tein]*v.*弄清;查明
underrate['ʌndə'reit]
*v.*低估;过低评价
budget['bʌdʒit]n.预算

glossing[glɔ:s]n.虚假外表;虚饰

---

[18] Activity[C, usually pl.]:a thing that you do in order to achieve a particular aim; （为达到一定目的而进行的）活动。
[19] OR:*(less formally)*Test data must be carefully prepared, results must be carefully reviewed, and corrections must be carefully made in the system.

It is essential that all aspects of testing be carefully considered and that they be as comprehensive as possible. To sure this, the development team works with users to devise a systematic *test plan*. The test plan includes the preparations for the series of tests previously described.

Conversion

Conversion is the process of changing from the old system to the new system. It answers the question, "Will the new system work under real conditions?" Four main conversion strategies can be employed:the *parallel strategy*, the *direct cutover strategy*, the *pilot study strategy*, and the *phased approach strategy*.

A formal conversion plan provides a schedule of all the activities required to install the new system. The most *time-consuming* activity in many cases is the conversion of data. Data from the old system must be transferred to the new system, either manually or through special conversion software programs. The converted data then must be carefully verified for *accuracy* and *completeness*.

Moving from an old system to a new one requires that end users be trained to use the new system. Detailed *documentation*[20] showing how the system works from both a technical and end-user standpoint is finalized[21] during conversion time for use in training and everyday operations. Lack of proper training and documentation because of time and cost constraints contributes to[22] system failure so this portion of the systems development process is very important.

Production and Maintenance

After the new system is installed and conversion is complete, the system is said to be in *production*. During this stage, the system will be periodically reviewed by both users and *technical specialists* to determine how well it has met its original objectives and to decide whether any revisions or modifications are in order. Changes in hardware, software, documentation, or procedures to a production system to correct

devise[di'vaiz]*v.*设计
systematic[ˌsistə'mætik]*adj.*系统化的;成体系的;条理化的

schedule['ʃedjuːəl]*n.*日程安排;工作计划

verify['verifai]*v.*核实;查对

---

[20] Documentation[U]:the documents that are required for sth, or that give evidence or proof of sth;必备文件；证明文件。

[21] Finalize:to complete the last part of a plan, project, etc.;把（计划、项目等）最后定下来；定案。

[22] Contribute to sth:to be one of the causes of sth;是…的原因之一。

errors, meet new requirements, or improve processing efficiency are termed maintenance.

Studies of maintenance have examined the amount of time required for various maintenance tasks. Approximately 20 percent of the time is devoted to debugging or correct emergency production problem; another 20 percent is concerned with changes in data, files, reports, hardware, or system software. But 60 percent of all maintenance work consists of making user enhancements, improving documentation, and recording system components for greater processing efficiency. The amount of work in the third category of maintenance problems could be reduced significantly through better systems analysis and design practices.

TERMINOLOGY

acceptance testing	processing function
analyst	production
business procedure	programming
control	requirement change
conversion	requirements analysis
data model	software system
direct cutover strategy	system analysis
documentation	system design
economic feasibility	system designer
feasibility study	system solution
information requirement	system testing
input	systems development
logical design	systems development cost
maintenance	systems development process
operational feasibility	systems failure
output	technical feasibility
parallel strategy	technical specialist
phased approach strategy	test plan
physical design	testing
pilot study strategy	unit testing

EXERCISES

4.1 Translate each of the following key terms:

a)analyst

b)requirements analysis

c)system design

d)unit testing

e)technical feasibility

f)systems development cost

g)programming

h)maintenance

4.2 Fill in the blanks in each of the following statements:

a)Systems development refers to systems analysis, systems design, _____, testing, conversion, and production and maintenance.

b)Systems analysis involves a _____ study to determine whether the solution is feasible, given the organization's resources and constraints.

c)_____ feasibility means to determine whether the benefits of the proposed solution outweigh the cost.

d)_____ is a detailed statement of the information needs that a new system must satisfy; identifies who needs what information, and when, where, and how the information is needed.

e)During the _____ stage, system specifications are translated into program code.

f)Testing is classified into three major types:_____, system testing, and acceptance testing.

g)A(n) _____ includes preparations for the series of tests to be performed on the system.

h)Four main conversion strategies are the parallel strategy, the direct cutover strategy, the _____strategy, and the phased approach strategy.

i)Systems design consists of _____ and physical design.

j)The most time-consuming activity in many cases is the conversion of _____.

k)After a new system is stalled and conversion is complete, the system is said to be in _____.

4.3 State whether each of the following is *true* or *false*. If *false*, explain why.

a)Although systems development activities usually take place in sequential order, some of them may need to be repeated or some may be taking place simultaneously.

b)From organizational analysis, system analysts are to detail the problems or limitations of existing sytems.

c)Feasibility study is of three major areas:technical, economic, and hardware feasibility study.

d)ASCII is the most widely used decimal code for computers.

e)The most difficult task of system analysts is to identify the information requirements that must be met by a system solution.

f)Logical design produces the actual specifications for hardware, software, physical databases, I/O media, manual procedures, and specific controls.

g)Programming constitutes a bigger portion of the systems developments cycle than design and the testing activities.

h)Systems development processes create requirement changes.

i)A conversion plan provides a schedule of all activities required to install the new system.

j)Documentation is the descriptions of how a new system works from both a technical and end-user standpoint.

4.4 Match each numbered item with the most closely related lettered item:

a)systems analysis	1. the exhaustive and thorough process that determines whether the system produces the desired results under known conditions.
b)systems design	2. changes in hardware, software, documentation, or procedures to a production system to correct errors, meet new requirements, or improve processing efficiency.
c)programming	3. the process of changing from the old system to the new system.
d)testing	4. details how a system will meet the information requirements as determined by the systems analysis.
e)conversion	5. the analysis of a problem that the organization will try to solve with a new system.
f)production	6. the process of translating the system specifications prepared during the design stage into program code.
g)maintenance	7. the stage after the new system is installed and the conversion is complete; during this time the system is reviewed by users and technical specialists to determine how well it has met its original goals.

4.5 What is the difference between systems analysis and systems design?

4.6 What is feasibility? Name and describe each of the three major areas of feasibility for softwre system.

4.7 What are information requirements? Why are they difficult to determine correctly?

4.8 Why is the testing stage of systems development so important? Name and describe the three stages of testing for an software system.

4.9 What roles do programming, production, and maintenance play in systems development?

Reading Material（阅读材料）

Client/Server and *N*-Tier Architectures

Distributed applications have come about through evolution rather than a revolution, albeit with occasional sudden leaps as new technology has become available. From the first hesitant decoupling of the desktop from the *database server* over a *low-bandwidth* network to the current situation in which the Internet predominates, the basic goal has remained the same—access and process shared data in a timely and secure manner.

Two-Tier Architecture

The classic *two-tier architecture* is what many developers mean by *client/server*, although if you put two developers in a room and ask them for a definition of client/server, you'll most likely end up with three opinions.

Here's how we define the terms *client* and *server*:

● A client is a process that requests a service (such as access to a database).

● A server is a process (such as a database server) that provides a server.

Notice that these definitions are *process-oriented*. It is perfectly feasible for both the client process and the server process to be located on the same machine, although in many cases they will be distributed—that's the more common case, especially when a single server is handling *requests* from *multiple clients*. The usual example of this sort of architecture is the *database management system(DBMS)*. Figure 4.2 depicts a typical database client/server system.

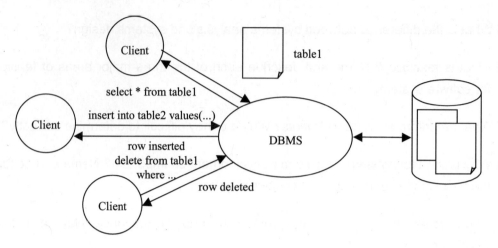

Figure 4.2   A database client/server system

In this case, the services being provided are requests to retrieve data and requests to create, modify, or remove items from the database. The client application and the server communicate using an agreed-upon mechanism—for example, *structured query language(SQL)* over a common *network protocol*. The server interprets the *SQL statement*, processes it, and sends an appropriate *response* back to the client. If the server is serving multiple clients simultaneously(using *multiple threads*), it might need to ensure that concurrent requests do not conflict. For example, two clients might attempt to update the same data at the same time; the server should reschedule one client, making it wait until the other has performed its work and released the necessary resources.

All of this is transparent to the client, in theory. In practice, however, a client might notice the delays that occur while the server tries to work its way through the competing client requests. The database design, application code, and other tuning issues become paramount in ensuring that the system scales well.

Further complications can arise when a designer realizes that a server can also be a client, and vice-verse; in order to perform its service, a server might need to request services from other servers, which in turn might use further servers, and so on. In a *distributed database*(depicted in Figure 4.3), the DBMS might be configured to forward requests to access data in a particular table to another DBMS(using a linked SQL Server, for example).

(Note: One could conceivably end up in a situation in which the DBMS that is contacted directly by a client simple becomes a channel through which all requests for data are passed to one or more other DBMSs. The original DBMS will still perform a useful service and can enforce various and integrity checks as the client requests or updates data, but it's not being very DBMS-like anymore.)

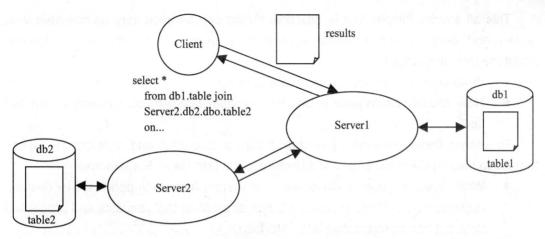

Figure 4.3   A distributed database (SQL Server)

N-Tier Architecture

N-tier systems(where *n* is 3 or more) require careful design and even more careful implementation. It is usual, but not mandatory, for each tier to contain components that perform well-defined roles. Different organizations give different names to each tier, reflecting the responsibilities of each tier. Examples of tiers from the Microsoft world include the *User Services Tier*, the *Business Services Tier*, and the *Data Services Tier*. The User Services Tier contains the presentation logic used for displaying data and gathering user input, the Business Services Tier. The User Services Tier contains the presentation logic used for displaying data and gathering user input, needed to manage persistent storage, usually using a DBMS.

Figure 4.4 depicts an example of a three-tier architecture.

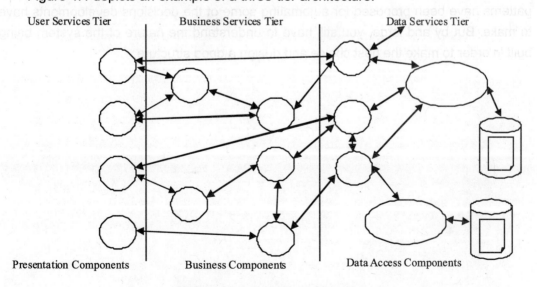

Figure 4.4   A three-tier architecture

This all sounds simple. But in practice, things can become very complicated very quickly, and there's a lot of room for developers to have deep and meaningful debates about the following issues:

- What code should actually go into each tier?
- How should components in one tier communicate with components in another tier?

To answer these questions, you should take a step back and consider the issues that led to the development of this architecture in the first place. For example

- What type, or types, of client will your system have to support: a GUI desktop application, an HTML browser, a batch application that requests and processes data but has no interactive user interface?
- Where is the data held, and in what format? Is there more than one data store?
- What are the data integrity requirements of the system? Where is the integrity of data checked? How do the business rules change over time?

You should also bear in mind that the tiers that result from partitioning a system in this manner are logical rather than physical. Many developers have been known to mistakenly assume that a three-tier architecture implies three separate computers: one for the desktop, another running the business components, and a third holding the database. This is not always so. Often, part of the *business logic* is located on the user's physical machine, some of the *data storage logic* might be located on the machine implementing the business rules, and the implementation of the business rules might be spread across several computers.

The choice of what components should go where is all part of the design. Many patterns have been proposed for automating some of the decisions developments have to make. But by and large, you still have to understand the nature of the system being built in order to make the best choice and design a good structure.

Note

Lesson 5　Data Structure Fundamentals

Software engineering is the study of ways in which to create large and complex computer applications and that generally involve many programmers and designers [1]. The heart of software engineering is with [2] the overall design of the applications and on [3] the creation of a design that is based on the needs and requirements of end users. While software engineering involves the full life cycle of a software project, it includes many different components—specification, requirements gathering, design, verification, coding, testing, quality assurance, user acceptance testing, production, and ongoing maintenance.

overall['əuvərɔːl]*adj.* 全面的;总体的

Having an in-depth understanding on every component of software engineering is not mandatory, however, it [4] is important to understand that the subject of data structures is concerned with the *coding phase*. The use of data structures is the nuts-and-blots [5] used [6] by programmers to store and manipulate data. This article briefly focuses on the essentials of data structures. Attempts will be made to understand how they work, which structure is best in a particular situation in an easy to understand environment.

mandatory['mændə ˌtɔːri]*adj.* 强制的

manipulate[mə'nipju leit]*v.*（熟练地）操作，使用

essential[i'senʃəl]*n.* 要点;要素

attempt[ə'tempt]*n.* 尝试;试图

A *data structure* is an arrangement of [7] data in a computer's memory or even disk storage. Examples of several common data structures are arrays, lists, queues, stacks, trees, graphs, and tables.

Arrays

An *array* is an object that consists of a sequence of numbered *elements*, all of the same type. The elements are numbered

---

[1] Tip:Here the relative pronoun "which" refers to "ways", and the relative pronoun 'that' also refers to "ways".

[2] With(prep.):concerning; in the case of;关于；对于。

[3] On(prep.):about sth;关于。

[4] It:used in the position of the subject or object of a verb when the real subject or object is at the end of sentence;用作形式主语或形式宾语，而真正的主语或宾语在句末。

[5] The nuts and bolts/the nuts-and-blots:the basic practical details of a subject or an activity;基本要点。

[6] OR:...the nuts-and-blots that are used...

[7] Of:used after nouns formed from verbs. The noun after "of" can be either the object or the subject of the action;用于由动词转化的名词之后，of 之后的名词可以是受动者，也可以是施动者。

beginning with 0 and can be referenced by their number using the *subscript operator [ ]*[8]. Arrays are widely used because they are so efficient. The length of an array is its number of elements.

reference['refrəns]*v.*
查阅;索引

Stacks

A *stack* is a container that implements the *last-in-first-out(LIFO)* protocol. This means that the only accessible object in the container is the last one among those that were inserted. A stack of books is a good analogy: you can't take a book from the stack without first removing the books that are stacked on top of it.

analogy[ə'nælədʒi]*n.*
类比;比喻

Queues

A *queue* is a container that implements the *first-in-first-out(FIFO)* protocol. That means that the only accessible object in the container is the one among them that was inserted first[9]. A good analogy is a group of people waiting[10] in line for a movie:the next one admitted is the person in the line who got there ahead of every one else.

Lists

A *list* is a sequential container that can insert and delete elements locally in constant time; i.e., at a rate that is independent of the size of the container. It is the preferred data structure for applications that do not need *random access*. A good analogy is a train of box cars:any car can be removed simply by disconnecting it from its two neighbours and then reconnecting them.

Trees

A *tree* is a *nonlinear* data structure that models a hierarchical organization. The characteristic features are that each element may have several *successor*(called its "*children*") and every element except one(called the "*root*") has a unique *predecessor*(called its "*parent*"). Trees are common in computer science:Computer file systems are trees, the inheritance structure for Java classes is a tree, the run-time system of method

hierarchical[ˌhaiə'ra: kikəl]*adj.*层次性的

inheritance[in'heritə ns]*n.*继承

---

[8] Tip:The element are numbered beginning with 0 and ... by their number using the subscript operator [ ]
[9] First(adv.):before anyone or anything else; at the beginning;首先。
[10] 同:...a group of people who are waiting...

invocation[ˌinvəˈkeɪʃən]*n.*调用

syntactical[sinˈtæktikəl]*adj.*句法的

invocations during the execution of a Java program is a tree, the classification of Java types is a tree, and the actual syntactical definition of the Java programming language itself forms a tree.

Tables

A *table*(also called a *map*, a *lookup table*, an *associative array*, or a *dictionary*) is a container that allows direct access by an *index type*. It works like an array except that the *index variable* need not be an integer. A good analogy is a dictionary; the index variable is the word being looked up, and the element that it indexes is its dictionary definition.

A table is a sequence of pairs. The first component of the pair is called the *key*. It serves as the index into the table, generalizing the subscript integer used in arrays. The second component is called the *value* of its key component. It contains the information being looked up[11]. In the dictionary example, the key is the word being[12] looked up, and the value is that word's definition(and everything else listed for that word).

A table is also called a map because we think of[13] the keys being mapped into their values, like a mathematical function: f(key)=value. Tables are also called an associative arrays because they can be implemented using two parallel arrays; the keys in one array and the values in the other.

Graphs

finite[ˈfainait]*adj.* 有 限的;有限制的

A *graph*[14] is a pair G=(V,E), where V and E are finite sets and every element of E is a two-element subset of V. The elements of V are called *vertices*[15] (or *nodes*), and the elements of E are called *edges*(or *arcs*). If e∈E then e={a,b} for some a,b∈V. In this case, we can denote e more simply as e=ab=ba. We say that the edge e connects the two vertices a and b, that e is incident with a and b, that a and b are incident upon e, that a and b are the *terminal points* or *endpoints* of the edge e, and that a and b are adjacent. The size of a graph is the number of

denote[diˈnəut]*v.*表示;标识

incident[ˈinsidənt]*adj.* 伴随而来的;自然的

adjacent[əˈdʒeisənt]*adj.*邻接的;毗邻的

---

[11] Look sth up:to search for information in a dictionary or another book, or by using a computer;（在词典、其他书中或通过计算机）查阅，检查。

[12] 同:...the word that is being looked up,...

[13] Of:used after some verbs before mentioning sb/sth involved in the action;用于某些动词后，后接动作所涉及的人或事。

[14] Stacks:栈; queues:队列;graphs:图。

[15] Vertex:a point where two lines meet to form a angle;顶点。

elements in its vertex set.

The data structures shown above(with the exception of the array) can be thought of as[16] Abstract Data Types. An *Abstract Data Type(ADT)* is more a way of looking at a data structure:focusing on what it does and ignoring how it does its job. A stack or a queue is an example of an ADT. It is important to understand that both stacks and queues can be implemented using an array. It is also possible to implement stacks and queues using a *linked list*. This demonstrates the "abstract" nature of stacks and queues:how they can be considered separately from their implementation. To best describe the term ADT, it is best to break the term down into "data type" and then "abstract".

When we consider a *primitive type* we are actually referring to two things:a *data item* with certain characteristics and the *permissible operations* on that data. An int in Java, for example, can contain any whole-number value from -2,147,483,648 to +2,147,483,647. It can also be used with the operators +, -, *, and /. The data type's permissible operations are an inseparable part of its identity; understanding the type means understanding what operations can be performed on it. In Java, any class represents a data type, in the sense that a class is made up of data(fileds) and permissible operations on that data(methods). By extension, when a data storage structure like a stack or queue is represented by a class, it too can be referred to as a data type. A stack is a different in many ways from an int, but they are both defined as a certain arrangement of data and a set of operations on the data.

Now let's look at the "abstract" portion of the phrase. The word "abstract" in our context stands for[17] "considered apart from the detailed specifications or implementation". In Java, an ADT is a class considered without regard to its implementation. It can be thought of as a "description" of the data in the class and a list of operations that can be carried out on that data and instructions on how to use these operations[18]. What is excluded

---

[16] Think of sb/sth as sb/sth:to consider sb/sth in a particular way;把…看作;把…视为。
[17] Stand for sth:to be an abbreviation or symbol of sth;（指缩写或符号）是…意思，代表。
[18] Tip:此句可简略成=It can be thought of as a "description" of the data ... and a list of operations ... and instructions ... .

though, is the details of how the methods carry out their tasks. An end user(or class user), you should be told[19] what methods to call, how to call them, and the results that should be expected, but not how they work.

We can further extend the meaning of the ADT when applying[20] it to data structures such as a stack and queue. In Java, as with any class, it means the data and the operations that can be performed on it. In this context, although, even the fundamentals of how the data is stored should be invisible to the user. Users not only should not know how the methods work, they should also not know what structures are being used to store the data. Consider for example the stack class. The end user knows that push() and pop()(among other similar methods) exist and how they work. The user doesn't and shouldn't have to know how push() and pop() work, or whether data is stored in an array, a linked list, or some other data structure like a tree.

The ADT specification is often called an interface. It's what the user of the class actually seen. In Java, this would often be the public methods. Consider for example, the stack class—the public methods push() and pop() and similar methods from the interface would be published to the end user.

TERMINOLOGY

Abstract Data Type(ADT)	lookup table
arc	map
array	node
associative array	nonlinear
children	parent
coding phase	permissible operation
data item	predecessor
data structure	primitive type
dictionary	queue

---

[19] Tell—[VNN]。
[20] 同:...when we apply ...

edge	random access
element	root
endpoint	stack
first-in-first-out(FIFO)	subscript operator []
graph	successor
index type	table
index variable	terminal point
key	tree
last-in-first-out(LIFO)	value
linked list	vertex
list	

EXERCISES

5.1 Translate each of the following key terms:

    a)data structure

    b)successor

    c)graph

    d)queue

    e)root

    f)stack

    g)linked list

    h)array

5.2 Fill in the blanks in each of the following statements:

    a)A(n) _____ is an arrangement of data in a computer's memory or even disk storage.

    b)The elements of an array are numbered beginning with _____.

    c)The length of an array is its number of _____.

    d)The LIFO protocol means that the only accessible object in a stack is the _____ one among those that were inserted.

    e)Every element of a tree except one, called the "_____", has a unique predecessor, called its "parent".

    f)In a graph, the elements of V are called _____.

    g)The size of a graph is the number of elements in its _____ set.

    h)The FIFO protocol means that the only accessible object in a queue is the _____ one among those that were inserted.

i)A table is also called a(n) _____.

j)An ADT focuses on _____ it does and ignores how it does its job.

5.3 Match each numbered item with the most closely related lettered item:

a)array	1.container that implements the FIFO protocol.
b)stack	2.object that consists of a sequence of numbered elements, all of the same type.
c)queue	3.pair G=(V,E) where V and E are finite sets and every element of E is a two-element subset of V.
d)list	4.nonlinear data structure that models a hierarchical organization.
e)tree	5.container that allows direct access by an index type.
f)table	6.container that implements the LIFO protocol.
g)graph	7.sequential container that can insert and delete elements locally in constant time.

5.4 Expand each of the following acronyms:

a)LIFO.

b)OOP.

c)ADT.

d)FIFO.

Reading Material （*阅读材料*）

Trees, Binary Trees, and Binary Search Trees

Linked lists usually provide greater flexibility than arrays, but they are linear structures and it is difficult to use them to organize a hierarchical representation of objects. Although stacks and queues reflect some hierarchy, they are limited to only one dimension. To overcome this limitation, we create a new data type called a *tree* that consists of *nodes* and *arcs*. Unlike natural trees, these trees are depicted upside down with the *root* at the top and the *leaves* at the bottom. The root is a node that has no parent; it can have only child nodes. Leaves, on the other hand, have no children, or rather, their children are empty structures. A tree can be defined recursively as the following:

1. An empty structure is an empty tree.

2. If $t_1, ..., t_k$ are disjoint trees, then the structure whose root has as its children the

roots of t_1, ..., t_k is also a tree.

3. Only structures generated by rules 1 and 2 are trees.

Figure 5.1 contains examples of trees. Each node has to be reachable from the root through a unique sequence of arcs, called a *path*. The number of arcs in a path is called the *length* of the path. The *level* of a node is the length of the path from the root to node plus 1, which is the number of nodes in the path. The *height* of a nonempty tree is the maximum level of a node in the tree. The empty tree is a legitimate tree of height 0(by definition), and a single node is a tree of height 1. This is the only case in which a node is both the root and a leaf. The level of a node must be between 1(the level of the root) and the height of the tree, which in the extreme case is the level of the only leaf in a degenerate tree resembling a linked list.

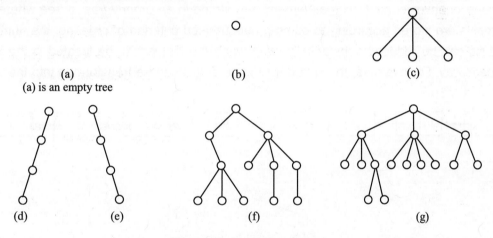

(a)
(a) is an empty tree
(b)
(c)
(d)
(e)
(f)
(g)

Figure 5.1   Examples of trees

Figure 5.2 contains an example of a tree that reflects the hierarchy of a university. Other examples are genealogical trees, trees reflecting the grammatical structure of sentences, and trees showing the taxonomic structure of organism, plants, or characters. Virtually all areas of science make use of trees to represent hierarchical structures.

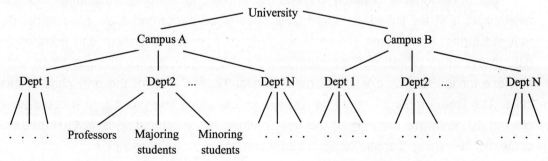

Figure 5.2   Hierarchical structure of a university shown as a tree

The definition of a tree does not impose any condition on the number of children of a given node. This number can vary from 0 to any integer. In hierarchical trees, this is a welcome property. For example, the university has only two branches, but each campus can have a different number of departments. Such trees are used in database management systems, especially in the hierarchical model. But representing hierarchies is not the only reason for using trees. In fact, in the discussion to follow, that aspect of trees is treated rather lightly, mainly in the discussion of expression trees.

Consider a linked list of n elements. To locate an element, the search has to start from the beginning of the list, and the list must be scanned until the element is found or the end of the list is reached. Even if the list is ordered, the search of the list always has to start from the first node. Thus, if the list has 10,000 nodes and the information in the last node is to be accessed, then all 9999 of its predecessors have to be traversed, an obvious inconvenience. If all the elements are stored in an *orderly tree*, a tree where all elements are stored according to some predetermined criterion of ordering, the number of tests can be reduced substantially even when the element to be located is the one further away. For example, the linked list Figure 5.3(a) can be transformed into the tree in Figure 5.3(b).

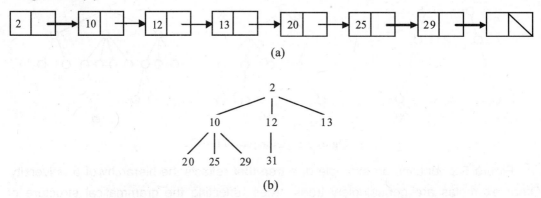

(a)

(b)

Figure 5.3   Transforming (a) a linked list into (b) a tree

Was a reasonable criterion of ordering applied to construct this tree? To test whether 31 is in the linked list, eight tests have to be performed. Can this number be reduced further if the same elements are ordered from top to bottom and from left to right in the tree? What would an algorithm be like that forces us to make three tests only:one for the root, 2, one for its middle child, 12, and one for the only child of this child, 31? The number 31 could be located on the same level as 12, or it could be a child of 10. With this ordering of the tree, nothing really interesting is achieved in the context of searching. Consequently, a better criterion must be chosen.

Again, note that each node can have any number of children. In fact, there are algorithms developed for trees with a deliberate number of children, but here discusses only binary trees. A *binary tree* is a tree whose nodes have two children(possibly empty), and each child is designated as either a left child or a right child. For example, the tree in Figure 5.4 are binary trees, whereas the university tree in Figure 5.2 is not. An important characteristic of binary trees, which is used in assessing an expected efficiency of sorting algorithms, is the number of leaves.

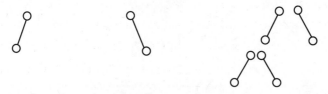

Figure 5.4   Examples of binary trees

As already defined, the level of a node is the number of arcs traversed from the root to the node puls one. According to this definition, the root is at level 1, its nonempty children are at level 2, and so on. If all the nodes at all levels except the last had two children, then there would be $1=2^0$ node at level 1, $2=2^1$ nodes at level 2, $4=2^2$ nodes at level 3, and generally, 2^i nodes at level i+1. A tree satisfying this condition is referred to as a *complete binary tree*. In this tree, all nonterminal nodes have both their children, and all leaves are at the same level. Consequently, in all binary trees, there are at most 2^i nodes at level i+1.

The *binary search trees*, also called *ordered binary trees*, are of particular interest. A binary search tree has the following property:for each node n of the tree, all values stored in its left subtree(the tree whose root is the left child) are less than value v stored in n, and all values stored in the right subtree are greater than v.(For reasons to be discussed, storing multiple copies of the same value in the same tree is avoided.) An attempt to do so can be treated as an error. The meanings of "less than" or "greater than" depend on the type of values stored in the tree. We use operators "<" and ">" which can be overloaded depending on the content. Alphabetical order is also used in the case of strings.

Note

Lesson 6   Object-Oriented Programming vs. Procedural Programming

In the early days of programming, programs were designed using *flowcharts* and a sort of *top-down design*[1]. With this type of design, a large problem is solved by breaking it down into[2] smaller tasks. For each of[3] the smaller tasks, a *procedure* is written. One *main procedure* was written to start the process and subsequently flow through to the solution, invoking[4] the desired procedures along the way.

flowchart['fləʊˌtʃɑːt]*n.* 流程图

This type of programming is referred to as *procedural programming*. There are many *procedural programming languages* widely used today, most notably COBOL and C.

Procedural programming involves writing a procedure that performs a specific task. Any data that the procedure needs to use is passed[5] in to the procedure via its *parameters*. The procedure can view and alter the data passed in and optionally return a value back to whoever called the procedure.

notably['nəʊtəbli]*adv.* 尤其;特别

parameter[pə'ræmitə] *n.*参数

To demonstrate procedural programming, let's look at an example. Suppose that you want to write a program to weekly pay the employees of a company(a realistic problem to solve). Paying employees involves computing their pay based on hours worked or a portion of an annual salary. In addition, you have to compute Social Security and Medicare taxes, as well as any income taxes to be withheld.

demonstrate['demən streit]*v.*示范;演示

Each of these computations has to be repeated for each employee in the company. Because these tasks are repeated, you can write a procedure for each one. For example, you might write a procedure called computePay() that inputs an employee's payment data and return his or her pay. You may also have procedures called computeMedicareTax(), computeSSTax(),

---

[1] Tip:..., programs were designed using flowcharts and a sort of top-down design.
[2] Break sth down into sth:to separate sth into smaller parts in order to analyse it or deal with it more easily;把…分类；划分。
[3] Of:used to show sb/sth belongs to a group, often after *some, a few*,etc.;（常用在 some、a few 等词语之后，表示人或物的所属）属于…的。
[4] 同:...to the solution, because the main procedure invoked...
[5] Pass sth in—[VN +adv.]。

and so forth. Each of these procedures needs to have the employee's data passed in to it.

occurrence[əˈkʌrəns]
n.事情

modify[ˈmɔdifait]v. 调整;稍作修改

approach[əˈprəutʃ]n.
方式;方法

This is a common occurrence[6] in procedural programming, in which data is passed around[7] between procedures. The procedures modify the data passed in and/or return back to whoever called the procedure. As you will soon see, object-oriented programming uses a different approach.

As your employee program evolves, you will certainly find[8] yourself adding and changing procedures so that everything works successfully. For example, as you start writing the computePay() procedure, you might realize that there are two different types of employees(at least in terms of how they get paid):hourly employees and salaried employees. In this situation, you might decide to write two computePay() procedures, one for hourly employees and a different one for salaried employees.

Writing a procedure to solve a specific task is a *fundamental programming* concept used in both procedural languages and in object-oriented programming. As you design and write Java programs, procedures are referred to as methods. Methods appear within a class. Procedures in a procedural language typically appear at a global level so that they can be invoked from anywhere.

Object-oriented programming(OOP) originated from research done by Xerox's Palo Alto Research Center(PARC) in the 1970s. OOP takes an entirely different approach to [9] developing computer applications. Instead of designing a program around the tasks that are to be solved, a program is designed around the objects in the problem domain.

You can think of procedural programming as writing a procedural for the verbs in the problem domain, such as paying an employee or computing taxes. You can think of object-oriented programming as writing a class for each of the nouns

---

[6] Occurrence:something that happens or exists;发生的事情；存在的事物。

[7] Pass sth around:to give sth to another person/thing, who gives it to sb/sth else, etc. until everyone has seen it;挨个传递某物。

[8] Find[VN -ing]:to discover sb/sth/yourself doing sth or in a particular situation, especially when this is unexpected;发现，发觉（处于某状态、在做某事）。

[9] Approach[C] to sth:a way of dealing with sb/sth; a way of doing or thinking about sth such as a problem or a task;（待人接物或思考问题的）方式，方法，态度。

in the problem domain.(Granted[10], this may be oversimplifying OOP, but you need to conceptually understand this important difference between OOP and procedural programming.)

Let's take another look at the example in which a program is to be written to pay employees of a company on a weekly basis. Instead of approaching this problem from the point of view of all the little tasks that need to be performed, such as computing an employee's pay and taxes, you begin by determining the objects in the problem domain.

An *object* is any person, thing, or entity that appears in the problem domain. In our example, we want to pay employees, so the employees are objects. The employees work for a company, so the company is another object. After further analysis, you might decide that the payroll department is an object. After you start writing the program, other objects will be discovered that were not apparent in the initial design.

After you have determined the objects in the problem, you write a *class* to describe the *attributes* and *behaviors* of each object. For example, we will need an Employee class that contains the attributes and behaviors of an employee.

The attributes of an Employee object will be what the employee "has", such as a name, address, employee number, Social Security number, and so on[11]. Each of these attributes will be represented by a *field* in the Employee class.

The behaviors of an Employee object are what the employee object "does"(or, more specifically, what we want the object to do). Employees do many things, but for our purposes, we want to be able to compute their pay and mail them a paycheck once a week. These desired behaviors become *methods* in the Employee class.

For each employee in the company, we would instantiate an Employee object. If we have 50 employees, we need 50 Employee objects. In memory, this would create 50 names, addresses, salaries, and so on. Each employee would be

oversimplify['əuvə'si mplifai]*v.* 陈述过于简略;说明过于简化
conceptually[kən'sep tjuəli]*adv.* 在概念上

initial[i'niʃəl]*adj.* 最初的;开始的;初期的
determine[di'tə:min]*v.* 确定;裁决

instantiate[in'stænʃie it]*v.* 实例化

---

[10] Granted(adv.):used to show that you accept that sth is true, often before you make another statement about it; (表示肯定，然后再作另一番表述) 不错，的确。

[11] And so on| and so forth:used at the end of a list to show that it continues in the same way;…等等。

distinguished by a *reference*, so we would need 50 references as well.

With object-oriented programming, data is still passed around between method calls as in procedural programs. However, there is an important distinction to be made when comparing procedural programming with object-oriented programming. The data that is passed around in an object-oriented program is typically varying data, such as the number of hours an employee has worked in a week. It is not the entire Employee object that gets passed around.

If a procedure in a procedural program needs data to perform a task, the necessary data is passed in to the procedure. With object-oriented programming, the object performs the task for you, and the method can access the necessary data without having to pass it in to the method.

For example, if you want to compute an employee's pay, you do not pass the corresponding Employee object to a computePay() method. Instead, you invoke the computePay() method on the desired Employee object. Because it is a part of the Employee object, the computePay() method has access to all the fields in the Employee object, including the Employee object's hourly pay, salary, and any other required data.

With object-oriented programming, an object's data are hidden from other parts of the program and can only be manipulated from inside the object. The method for manipulating the object's data can be changed internally without affecting other parts of the program. Programmers can focus on what they want an object to do, and the object decides how to do it.

An object's data are encapsulated from other parts of the system, so each object is an independent *software building block* that can be used in many different systems without changing the program code. Thus, objected-oriented programming is expected to reduce the time and cost of writing software by producing *reusable* program code or software *chips* that can be reused in other related systems.

---

distinction[disˈtiŋkʃən]*n.*差别;区分

varying[ˈveəriŋ]*adj.* 变化着的

encapsulate[enˈkæpsəˌleit]*v.*封装

TERMINOLOGY

attribute

behavior

chip

class

field

flowchart

fundamental programming

main procedure

method

object

object-oriented programming(OOP)

parameter

procedural programming

procedural programming language

procedure

reference

reusable

software building block

top-down design

EXERCISES

6.1 Translate each of the following key terms:

a)procedural programming

b)object-oriented programming

c)object

d)class

e)reusable

f)attribute

g)behavior

h)method

6.2 Fill in the blanks in each of the following statements:

a)_____ programming is a traditional software development method that has treated data and procedures as independent components.

b)In procedural languages, one _____ procedure is written to start a process.

c)_____ programming is an approach to software development that combines data and procedures into a single object.

d)Procedures are termed _____ in object-oriented languages.

e)Writing a procedure to solve a specific task is a _____ programming concept.

f)A class provides methods of a special type, known as _____, which are invoked to construct objects from the class.

6.3 Match each numbered item with the most closely related lettered item:

a)OOP	1.construct that defines objects of the same type.
b)OO software development	2.programming language designed to express the logic that can solve problems.
c)Procedural language	3.anyone that appears in a problem domain.
d)Object	4.methodology in which a program is organized into objects, each containing both the data and processing operations necessary to perform a task.
e)Class	5.software development approach that focuses less on the tasks and more on defining the relationships between previously defined procedures or objects.
f)Method	6.collection of statements that are group together to perform an operation.

6.4 Categorize each of the following programming languages as either procedural lanuages or object-oriented languages:

a)FORTRAN.
b)COBOL.
c)Pascal.
d)C++.
e)C#.
f)Visual Basic.
g)C.
h)Java.
i)Ada.
j)Baisc.

6.5 Expand each of the following acronyms:

a)OOP.
b)PARC.
c)OO.
d)OT.
e)DBMS.

Reading Material（阅读材料）

Basic Ideas and Benefits of Object Technology

Changes to data structures account for around 16% of Object Technology spending. The phrase *Object Technology(OT)* refers to several things, as do the terms *Object-Oriented(OO)* and *Object-Oriented methods* themselves. In particular, the phrase refers to objected-oriented programming, design, analysis and databases, in fact to a whole philosophy of systems development and knowledge representation based on a powerful metaphor.

In order to understand the basic of OT let us try to understand why this is so for conventional computer systems and see how OT helps to reduce the burden if properly applied. In doing so we will begin to grasp the meaning of the fundamental terms.

Being based on the so-called *Von Neumann architecture* of the underlying hardware, a conventional computer system can be regarded as a set of functions or processes together with a separate collection of data; whether stored in memory or on disk does not matter. This static *architectural model* is illustrated in Figure 6.1 which also indicates that, when the system runs, the dynamics may be regarded as some function, f(1), reading some data, A, transforming them and writing to B. Then some other function, f(2), reads some data, perhaps the same data, does whatever it does and writes data to C. Such overlapping data access gives rise to complex *concurrency* and *integrity* problems but these can be solved well by using a *database management system*. The question that I ask you to consider before reading on is:what must be done when part of the data structure has to change?

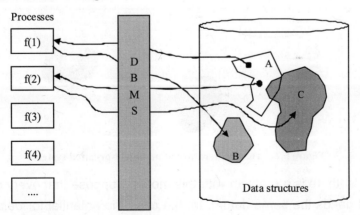

Figure 6.1   The architecture of a conventional computer system

Considering this from the point of view of a maintenance programmer, the only conclusion that one can come to is that every single function must be checked to see if it may be destabilized by the change. Good *documentation* can help with this but is rarely available in practice. Part of the reason for this is that good documentation for this task would itself consist in an object-oriented description of the system and is unlikely to be divorced from an object-oriented implementation or at least design. Furthermore, every function that is changed to reflect the new structure may have side effects in other parts of the system. Perhaps this accounts for the extraordinarily high *costs of maintenance*.

Figure 6.2 illustrates a completely different architectural approach to systems. Here, all the data that a function needs to access are encapsulated with it in packages called *objects* and in such a way that the functions of no other object may access these data. Using a simile suggested by Steve Cook, these objects may be regarded as eggs. The yolk is their *data structure*, the white consists of the functions that access these data and the shell represents the signature of the publicly visible operations. The shell interface hides the implementation of both the functions and the data structures. Now suppose again that a data structure is changed in the egg 'shelled' for maintenance in Figure 6.2. Now the maintenance programmer need only check the white of this particular egg for the impact of the change; maintenance is localized. If the implementation changes, no other object can possibly be affected. This is *encapsulation*:data and processes are *combined* and *hidden* behind an *interface*.

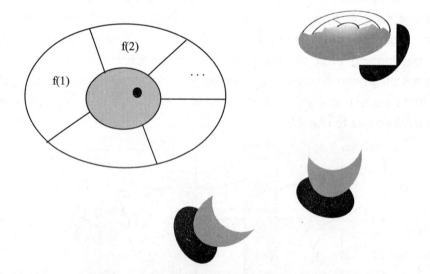

Figure 6.2   The arcliterature of an object-oriented system

As it stands, there is a problem with this model. Suppose that every object contains a function that needs the same datum. In that case the potential for *data duplication* is staggering and the approach would be quite impracticable. The solution is to permit the objects to send messages to each other. In this way object X may need data A but not encapsulate them. Provided that X stores in its yolk the identity of another object, Y, that does contain the required data, it may send a message requesting the data or even some transformed version of them. This is depicted in Figure 6.3, where the small sperm-like dot represents the identity of a target object and the arrows show the outward direction of the message. This, in a nutshell, is 50% of the idea behind object technology. The other 50% involves allowing the objects to be classified and related to each other in other ways. Notice that with this approach the maintenance problem is localized and thus greatly *simplified*. When a data structure changes, the maintainer need only check the functions in the albumen that encapsulates it. There can be no effect elsewhere in the system unless the shell is cracked or deformed; that is, if the interface changes. Thus, while we may claim to have reduced the maintenance problem by orders of magnitude, we now have to work very hard to ensure that we produce correct, complete and stable interfaces for our objects. This implies that sound analysis and design are even more worthwhile and necessary than they were for conventional systems. This extra effort is worthwhile because object technology leads to some very significant benefits.

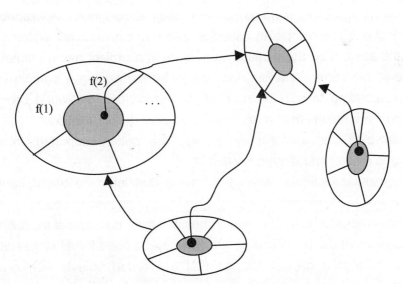

Figure 6.3   **Message passing eliminates data duplication**

These benefits are of much the same general character as those offered by structured programming and design, but go much further in some directions and, in doing so, lead to the questioning of many of the basic assumptions of the structured methods school of thought.

Anticipating our analysis but drawing on the remarks about information hiding and inheritance, the principle benefits are as follows.

- Maintenance is localized and thus is less costly and error-prone, even in the face of changing requirements, provided that the inheritance structure does not have to be rewritten.

- Object technology addresses the trade-off between quality and productivity. Well-designed object-oriented systems are the basis for systems to be assembled largely from *reusable components*, leading to higher productivity. Reusing *existing classes* that have been tested in the field on earlier projects leads to systems which are of higher quality, meet business requirements better and contain fewer *bugs*. This is probably the most publicized benefit of object technology.

- Object-oriented programming, and inheritance in particular, makes it possible to define clearly and use modules that are functionally incomplete and then allow their extension without upsetting the operation of other modules or their clients. This makes systems more *flexible*, more easily *extensible* and less costly to maintain.

- Object-orientation is a tool for managing *complexity*, leading to increased *scalability*. Partitioning systems on the basis of objects helps with the problem of scalability. There is no reason why effort should increase exponentially with project size and complexity as is the case with conventional systems.

- In the same way, the partitioning of work in a project has a natural basis and should be easier. Analysis and design by partitioning a domain into objects corresponding to a *solution-oriented* view of their real-world counterparts is often more natural than a *top-down* functional decomposition.

- Prototyping and evolutionary delivery are better supported, thus reducing time-to market and requirements drift.

- The message passing metaphor means that interface descriptions between modules and external and legacy systems become much easier.

- There is greater *seamlessness* in passing from *conceptual modelling*, through analysis and design to coding. Objects can be used for all stages of modelling and there is a greater change that the coded objects will correspond to something in the vocabulary of users:opening the possibility of shared understanding between developers and their clients.

- Object-oriented systems are potentially capable of capturing far more of the meaning of an application; its semantics. Since object-oriented methods are main concerned with modelling systems they can used to carry out scenario modelling and facilitate changes within the business. This property leads to greater reversibility in the end product and enhances the possibility of reverse engineering systems and of tracing features back to requirements.

- *Information hiding* through encapsulation helps to build secure systems.
- Formal specification methods can be made to blend with object-oriented design to some extent.
- Some applications have defeated other approaches and object technology seems to be the only way to build them efficiently. Examples are *graphical user interfaces*, *distributed systems*, *agent-based systems* and *workflow systems*.

Note

Lesson 7　Java

Java is a *high-level programming language* that[1] was developed by a team led by James Gosling at Sun Microsystems. Originally called *Oak*, it was designed in 1991 for use in *embedded consumer electronic appliances*. In 1995, named[2] *Java*, it was redesigned for developing *Internet applications*.

Java is a full-featured, general-purpose programming language that is capable of developing robust mission-critical applications. In recent years, Java has gained enormous popularity and has quickly become the most popular and successful programming language. Today, it is used not only for *Web programming*, but also[3] for developing *standalone applications* across platforms on *servers*, *desktops*, and *mobile devices*. It was used to develop the code to communicate with and control the robotic[4] rover that rolled on Mars. Many companies that once[5] considered Java to be more hype than substance are now using it to create *distributed applications* accessed by customers and partners across *the Internet*. For every new project being developed today, companies are asking how they can use Java to make[6] their work easier.

Java initially became attractive because Java programs can be run from[7] a *Web browser*. Java programs that run from a Web browser are called *applets*. Applets use a modern *graphical user interface* with buttons, text fields, text areas, radio buttons, and so on, to interact with users on the Web and process their requests. Applets make the Web responsive, interactive, and fun to use. To run applets from a Web browser, you need to use Netscape7 or Internet Explorer6, or higher.

Java can also be used to develop applications on the server

robust[rəu'bʌst]*adj.* 强壮的;强劲的

rover['rəuvə]*n.*漫游者
hype[haip]*adj.* 言过其实的;大肆宣传的

attractive[ə'træktiv] *adj.*（事物）有吸引力的;吸引人的

---

[1] Tip:Here the relative pronoun "that" refers to "Java", and the subject of the relative clause is also "Java". In this case, the relative pronoun can't be omitted.
[2] Name—[VN-N]。
[3] Not only...but (also)...:both...and...;不但…而且…。
[4] Robotic(adj.):connected with robots;机器人的。
[5] Once(adv.):at some time in the past;曾；曾经。
[6] Make—[VN-ADJ]。
[7] From:used to show what the origin of sb/sth is;（表示来源）来自，从…来。

side. These applications, called *Java servlets* or *JavaServer Pages(JSP)*, can be run from a Web server to generate *dynamic Web pages*.

versatile['vɔ:sɔtail]*adj*.多功能的;多用途的

Java is a versatile programming language. You can use it to develop applications on your desktop and on the server. You can also use it to develop applications for small *hand-held devices*, such as personal digital assistants and cell phones.

Java has become enormously popular. Java's rapid rise and wide acceptance can be traced to its design and programming features, particularly its promise that you can write a program once and run it anywhere. As stated in the *Java-language white paper* by Sun, Java is *simple*, *object-oriented*, *distributed*, *interpreted*, *robust*, *secure*, *architecture-neutral*, *portable*, *multithreaded*, and *dynamic*. Let's analyse

buzzword['bʌz,wɔ:d]*n*.（出版物上的）时髦用语,流行行话

these often-used buzzwords.

Simple

No language is simple, but Java is a bit[8] easier than the popular object-oriented programming language C++, which was the dominant software-development language before Java.

model['mɔdɔl]*v*. 复制;仿制

Java is partially modeled on C++, but greatly simplified and improved. For instance, *pointers* and *multiple inheritance* often make[9] programming complicated. Java replaces[10] the multiple inheritance in C++ with a simple language construct called an *interface*, and eliminates pointers.

construct[kɔn'strʌkt]*n*.构件;结构体
eliminate[i'limineit]*v*.排除;去除

Java uses *automatic memory allocation* and *garbage collection*, whereas C++ requires the programmer to allocate memory and collect garbage. Also, the number of language constructs is small for such a powerful language. The clean

syntax['sin,tæks]*n*.（计算机上的）句法;语构
negative['negɔtiv]*adj*.消极的;负面的

syntax makes Java programs easy to write and read. Some people refer to Java as "++--" because it is like C++ but with more functionality and fewer negative aspects.

Object-Oriented

inherently[in'hiɔrɔntli]*adv*.天生地;固有地

Java is inherently object-oriented. Although many object-oriented languages began strictly as *procedural language*, Java was designed from the start to be object-oriented. *Object-oriented*

---

[8] A bit(adv.):rather;有点儿；稍微。
[9] Make—[VN -ADJ]。
[10] Replace sb/sth by/with sb/sth:to remove sb/sth and put another person or thing in their place;（用…）替换；（以…）代替。

programming(OOP) is a popular programming approach that is replacing traditional *procedural programming* techniques.

Software systems developed using procedural programming languages are based on the paradigm of *procedures*. Object-oriented programming models the real world in terms of *objects*. Everything in the world can be modeled as an object. A circle is an object, a person is an object, and a Windows icon is an object. Even a loan can be perceived as an object. A Java program is object-oriented because programming in Java is centered on creating objects, manipulating objects and making[11] objects work together.

paradigm['pærə,daim] *n.*范例;样式

One of the central issues in software development is how to reuse code. Object-oriented programming provides great flexibility, modularity, clarity, and reusability through *encapsulation*, *inheritance*, and *polymorphism*[12]. For years, object-oriented technology was perceived as elitist, requiring a substantial investment in training and infrastructure. Java has helped[13] object-oriented technology enter the mainstream of computing. Its simple, clean syntax makes programs easy to write and read. Java programs are quite expressive in terms of designing and developing applications.

modularity[,mɔdju'læriti]*n.*模块化

elitist[ei'li:tist]*n.*杰出物;精品

Distributed

Distributed computing involves several computers working together on a network. Java is designed to make distributed computing easy. Since networking capacity is inherently integrated into Java, writing network programs is like sending and receiving data to and from a file.

Interpreted

You need an *interpreter* to run Java programs. The programs are compiled into the Java Virtual Machine code called *bytecode*. The bytecode is machine-independent and can run on any machine that has a Java interpreter, which is part[14] of the *Java Virtual Machine(JVM)*.

---

[11] Make—[VN inf]。
[12] Encapsulation:封装; inheritance:继承; polymorphism:多态。
[13] Help—[VN **to** inf]。
[14] Part$_{[U]}$ of sth:some but not all of a thing;部分。

Most *compilers*, including C++ compilers, translate programs in a high-level language to *machine code*. The code can only run on the *native machine*. For instance, if you compile a C++ program in Windows, the executable code generated[15] by the compiler can only run on the Windows platform. With Java, you compile the source code once, and the bytecode generated by a Java compiler can run on any platform with a Java interpreter. The Java interpreter translates the bytecode into the machine language of the target machine.

Robust

Robust means *reliable*. No programming language can ensure complete reliability. Java puts a lot of emphasis on early checking for possible error, because Java compilers can detect many problems that would first show up at execution time in other languages. Java has eliminated certain types of error-prone programming constructs found in other languages. It does not support pointers, for example, thereby[16] eliminating the possibility of overwriting memory and corrupting data.

Java has a *runtime exception-handling* feature to provide programming support for robustness. Java forces the programmer to write the code to deal with *exceptions*. Java can catch and respond to an exceptional situation so that the program can continue its normal execution and terminate gracefully when a *runtime error* occurs.

Secure

As an Internet programming language, Java is used in a networked and distributed environment. If you download a Java applet and run it on your computer, it will not damage your system because Java implements several *security mechanisms* to protect[17] your system against harm caused by stray programs. The security is based on the premise that nothing should be trusted.

Architecture-Neutral

Java is interpreted. This feature enables Java to be architecture-

detect[di'tekt]*v.* 发现；查出

corrupt[kə'rʌpt]*v.* 引起（计算机文件、数据等的）错误;破坏

exception[ik'sepʃən]*n.* 异常;例外的事物

terminate['tə:mineit]*v.* 结束;终结

harm[hɑ:m]*n.* 危害;损害

stray[strei]*adj.* 不守规矩的;偏离的

premise['premis]*n.* 前提;假定

---

[15] Tip:... the executable code generated by the compiler can... .
[16] Thereby(adv.):used to introduce the result of the action or situation mentioned;因此；从而。
[17] Protect sb/sth against/from sth:to make sure that sb/sth is not harmed, injured, damaged, etc.;保护；防护。

neutral, or to use an alternative term, *platform-independent.* With a JVM, you can write one program that will run on any platform.

> neutral['nju:trəl]*adj.*
> 中立的;无倾向性的

Java's initial success stemmed from its Web-programming capability. You can run Java applets from a Web browser, but Java is for more than just writing Web applets. You can also run stand-alone Java applications directly from operating systems, using a Java interpreter. Today, software vendors usually develop multiple versions of the same product to run on different platforms(Windows, OS/2, Macintosh, and various UNIX, IBM AS/400, and IBM mainframes). Using Java developers need to write only one version that can run on every platform.

> stem[stem]*v.*抑制;遏制

Portable

Because Java is architecture-neutral , Java programs are portable. They can be run on any platform without being recompiled. Moreover, there are no platform-specific features in the Java language. In some languages, such as Ada, the largest integer varies on different platforms. But in Java, the range of the integer is the same on every platform, as[18] is the behavior of arithmetic. The fixed rang of the numbers makes the program portable.

> arithmetic[ə'riθmətik]
> *n.*算术;计算

The Java environment is also portable to new hardware and operating systems. In fact, the Java compiler itself is written in Java.

Multithreaded

Multithreading is a program's capability to perform several tasks simultaneously. For example, downloading a video file while playing the video would be considered[19] multithreading. Multithreaded programming is smoothly integrated in Java, whereas[20] in other languages you have to call[21] procedures specific to the operating system to enable multithreading.

> simultaneously[saim
> əl'teiniəsli]*adv.*同时进
> 行地;同步地

Multithreading is particularly useful in graphical user interface(GUI) and network programming. In GUI programming,

[18] The same as...:having the same quality, number, size, etc.;（质量、数量、大小等）相同，一样。
[19] Consider—[VN -N]。
[20] Whereas:used to compare or contrast two facts;（用以比较或对比两个事实）然而，但是，尽管。
[21] Call:调用。

there are many things going on at the same time. A user can listen to an audio recording while surfing a Web page. In network programming, a server can serve multiple clients at the same time. Multithreading is a necessity in multimedia and network programming.

Dynamic

Java was designed to adapt[22] to an evolving environment. New classes can be loaded on the fly without recompilation. There is no need for developers to create, and for users to install, major new *software versions*. New features can be incorporated transparently as need.

transparently[træns'pɛərənt]*adv.* 透明地;方便易懂地

TERMINOLOGY

applet	mobile device
architecture-neutral	multiple inheritance
automatic memory allocation	multithreaded
bytecode	multithreading
compiler	native machine
desktop	Oak
distributed	object
distributed application	object-oriented
distributed computing	object-oriented programming(OOP)
dynamic	platform-independent
dynamic Web page	pointer
embedded consumer electronic appliance	polymorphism
encapsulation	portable
exception	procedural language
garbage collection	procedural programming
graphical user interface	procedure
hand-held device	reliable
high-level programming language	robust

---

[22] Adapt to sth:to change your behavior in order to deal more successfully with a new situation;适应（新情况）。

inheritance
interface
Internet application
interpreted
interpreter
Java
Java servlet
Java Virtual Machine(JVM)
Java-language white paper
JavaServer Page(JSP)
machine code

runtime error
runtime exception-handling
secure
security mechanism
server
simple
software version
standalone application
the Internet
Web browser
Web programming

EXERCISES

7.1 Translate each of the following key terms:

a)JavaServer Page
b)polymorphism
c)interpreter
d)applet
e)garbage collection
f)dynamic Web page
g)platform-independent
h)Java servlet

7.2 Fill in the blanks in each of the following statements:

a)Java was developed by a team led by James Gosling at _____.
b)Java programs that run from a Web browser are called _____.
c)_____ can be run from a Web server to generate dynamic Web pages.
d)Java has a runtime _____ feature to provide programming support for robustness.
e)_____ involves several computers working together on a network.
f)Object-oriented programming provides great flexibility, modularity, clarity, and reusability through encapsulation, _____, and polymorphism.
g)_____ is a program's capability to perform several tasks simultaneously.
h)Java uses _____ to allocate memory.
i)Java replaces the multiple inheritance in C++ with a simple language construct

called a(n) _____.

j)Robustness means _____.

k)Java is architecture-neutral, also called _____.

l)Java promise that you can write a program once and run it _____.

7.3 State whether each of the following is *true* or *false*. If *false*, explain why.

a)The Java interpreter is part of the JVM.

b)The Java compiler itself is written in C.

c)Multithreading is a necessity in multimedia and network programming.

d)The Java interpreter is capable of translating the machine language of a target machine into the bytecode.

e)The dynamic feature means that new classes can be loaded on the fly without recomplication.

f)C++ compilers translate C++ programs into bytecode.

g)Java is a programming language that can deliver only the software functionality needed for a particular task as a small applet down from a network; can run on any computer and operating system.

7.4 Expand each of the following acronyms:

a)JSP.

b)GUI.

c)JVM.

d)OS.

e)OOP.

f)IDE.

g)PC.

Reading Material（阅读材料）

Elementary C++ Programming

A *program* is a sequence of instructions that can be executed by a computer. Every program is written in some programming language. C++(pronounced "see-plus-plus") is one of the most powerful programming language available. It gives the programmer the power to write efficient, structured, object-oriented programs.

To write and run C++ programs, you need to have a text editor and a C++ compiler installed on your computer. A *text editor* is a software system that allow you to create and edit text files on your computer. Programmers use text editors to write programs in a programming language such as C++. A *compiler* is a software system that translates programs into the machine language(also called *binary code*) that the computer's operating system can then run. That translation process is called *compiling* the program. A *C++ compiler* compiles C++ programs into machine language.

If your computer is running a version of the Microsoft Windows operating system(e.g., Windows 98 or Windows 2000), then it already has two text editors: WordPad and Notepad. These can be stared from the Start key. In Windows 98, they are listed under Accessories.

Windows does not come with a built-in C++ compiler. So unless someone has installed a C++ compiler on the machine you are using, you will have to do that yourself. If you are using a Windows computer that is maintained by someone else(e.g., an Information Services department at your school or company), you may find a C++ compiler already installed. Use the Start key to look under Programs for Borland C++ Builder, Metrowerks CodeWarrior, Microsoft Visual C++, or any other program with "C++" in its name. These are usually referred to as *IDEs(Integrated Development Environments)* because they include their own specialized text editors and debuggers.

If your computer is running a proprietary version of the UNIX operating system on a workstation(e.g., Sun Solaris on a SPARC station), it may already have a C++ compiler installed.

Now you have a text editor for writing C++ programs and a C++ compiler for compiling them. If you are using an IDE such as Borland C++ Builder on a PC, then you can compile and run your programs by clicking on the appropriate buttons. Other systems may require you to use the command line to run your programs. In that case, you do so by entering the file name as a command. For example, if your source code is in a file named hello.cpp, type

> Hello

as the command line to run the program after it has been compiled.

When writing C++ programs, remember that C++ is *case-sensitive*. That means that main() is different from Main(). The safest policy is to type everything in lower-case except when you have a compelling reason to capitalize something.

EXAMPLE 7.1　The "Hello, World" Program

```
#include <iostream>
int main()
{ std::cout << "Hello,World!\n";
}
```

The first line of this source code is a *preprocessor directive* that tells the compiler where to find the definition of the std::cout object that is used on the third line. The identifier is the name of a file int the *Standard C++ Library*. Every C++ program that has standard input and output must include this preprocessor directive. Note the required punctuation: the pound sign # is required to indicate that the word "include" is a preprocessor directive; the angle brackets <> are required to indicate that the word "iostream"(which stand for "input/output stream") is the name of a Standard C++ Library file. The expression <iostream> is called a *standard header*.

The second line is also required in every C++ program. It tells where the program begins. The identifier main is the name of a function, called *the main function* of the program. Every C++ program must have one and only one main() function. The required parentheses that follow the word indicate that it is a function. The keyword int is the name of a data type in C++. It stands for "integer". It is used here to indicate the *return type* for the main() function. When the program has finfished running, it can return an integer value to the operating system to signal some resulting status.

The last two lines constitute the actual body of the program. A *program body* is a sequence of program statements enclosed in braces { }. In this example there is only one statement:

```
std::cout << "Hello,World!\n";
```

It says to send the string "Hello,World!\n" to the *standard output stream* object std::cout. The single symbol << represents the *C++ output operator*. When this statement executes, the characters enclosed in quotation marks " " are sent to the standard output device which is usually the computer screen. The last two characters \n represent the *newline character*. When the output device encounters that character, it advances to the beginning of the next line of text on the screen. Finally, note that every program statement must end with a semicolon(;).

Notice how the program in Example 1.1 is formatted in four lines of source code. That formatting makes the code easier for humans to read. The C++ compiler ignores formatting. It reads the program the same as if it were written all on one line, like this:

```
#include <iostream>
int main(){ std::cout << "Hello,World!\n";}
```

Blank spaces are ignored by the compiler except where needed to separate identifiers, as in

```
int main
```

Note that the preprocessor directive must precede the program on a separate line.

EXAMPLE 7.2 Another "Hello, World" Program

```
#include <iostream>
using namespace std;
int main()
```

```
{ //prints "Hello,World!"
    cout << "Hello,World!\n";
    return 0;
}
```

The second line

```
using namespace std;
```

tells the C++ compiler to apply the prefix std:: to resolve names that need prefixes. It allow us to use cout in place of std::cout. This makes larger programs easier to read.

The fourth line

```
{ //prints "Hello,World!"
```

includes the comment "prints "Hello,World!"". A *comment* in a program is a string of characters that the preprocessor removes before the compiler compiles the programs. It is included to add explanations for human readers. In C++, any text that follows the double slash symbol //, up to the end of the line, is a comment. You can also use C style comments, like this:

```
{ /*prints "Hello,World!"*/
```

A *C style comment* (introduced by the programming language named "C") is any string of characters between the symbol /* and the symbol */. These comments can run over several lines.

The sixth line

```
return 0;
```

is optional for the main() function in Standard C++. We include it here only because some compilers except it to be included as the last line of the main() function.

A *namespace* is a named group of definitions. When objects that are defined within a namespace are used outside of that namespace, either their names must be prefixed with the name of the namespace or they must be in a block that is preceded by a using namespace statement. Namespaces make it possible for a program to use different objects with the same name, just as different people can have the same name. The cout object is defined within a namespace named std (for "standard") in the <iostream> header file.

Note

Lesson 8   Overview of a Database Management System

Databases today are essential to every business. Whenever[1] you visit a major Web site—Google, Yahoo!, Amazon.com, or thousands of smaller sites that[2] provide information—there is a database behind the scenes serving up[3] the information you request[4]. Corporations maintain all their important records in databases. Databases are likewise found at the core of many scientific investigations. They represent the data gathered by astronomers, by investigator of the human genome, and by biochemists exploring properties of proteins, among many other scientific activities.

scene[si:n]n.情景；场景
corporation[ˌkɔ:pəˈreiʃən]n.（大）公司
investigation[inˌvestiˈgeiʃən]n.科学研究；学术研究

What is a database? In essence[5] a database is nothing more than a collection of information that exists over a long period of time, often many years. In common parlance, the term *database* refers to a collection of data that is managed by a *database management system*, or *DBMS*, or more colloquially a "*database system*". The DBMS is expected to:

parlance[ˈpɑ:ləns]n.说法；用语
colloquially[kəˈləukwi:əl]adv.口语化；口头化（地）

1. Allow users to create new databases and specify their *schemas*(logical structure of the data), using a specialized data-definition language.

2. Give[6] users the ability to *query* the data(a "query" is a database lingo for a question about the data) and *modify* the data, using an appropriate language, often called a *query language* or *data-manipulation language*.

lingo[ˈliŋgəu]n.术语；行话

3. Support the storage of[7] very large amounts of data—many terabytes[8] or more—over a long period of time, allowing efficient access to the data for queries and

---

[1] Whenever(conj.):at any time that; on any occasion that;在任何的时候；不论何时。

[2] Tip:Here the relative pronoun "that" refers to "thousands of smaller sites", and the subject of the relative clause is also "thousands of smaller sites". In this case, the relative pronoun can't be omitted.

[3] Serve sth up:to give, offer or provide sth;给出；提供。

[4] Tip:... a database behind the scenes serving up the information you request.

[5] In essence[U]:when you consider the most important points;从本质上讲。

[6] Give—[VNN]。

[7] Of:used after nouns formed from verbs. The noun after "of" can be either the object or the subject of the action;用于由动词转化的名词之后，of 之后的名词可以是受动者，也可以是施动者。

[8] Terabyte (abbr. TB):a unit of information equal to one million million, or 10^{12} bytes;万亿字节；太字节。

database modifications.

4. Enable *durability*, the recovery of the database in the face of failures, errors of many kinds, or intentional misuse.

5. Control access to data from many users at once, without allowing unexpected interactions among users(called *isolation*) and without actions on the data to be performed partially but not completely(called *atomicity*).

Figure 8.1 is an outline of a complete DBMS. Single boxes represent *system components*, while double boxes represent in-memory *data structures*. The solid lines indicate *control* and *data flow*, while dashed lines indicate data flow only. Since the diagram is complicated, we shall consider the details in several stages. First, at the top, we suggest that there are two distinct sources of commands to the DBMS:

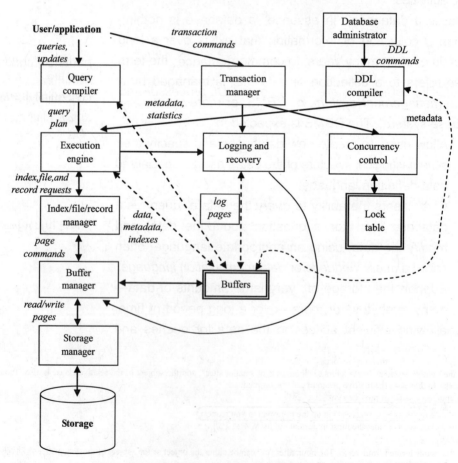

Figure 8.1   Database management system components

1.　Conventional users and application programs that ask for[9] data or modify data.

2.　A *database administrator*:a person or persons responsible[10] for the structure or schema of the database.

Data-Definition Language Commands

The second kind of command is the simpler to process, and we show its trail beginning at the upper right side of Figure 8.1[11]. For example, the database administrator, or *DBA*, for a university registrar's database might decide that there should be a *table* or *relation* with *columns* for a student, a course the student has taken, and a grade for that student in that course. The DBA might also decide that the only allowable grades are A, B, C, D, and F. This structure and constraint information is all part of the schema of the database. It is shown in Figure 8.1 as entered by the DBA, who needs special authority to execute *schema-altering commands*, since[12] these can have profound effects on the database. These schema-altering *data-definition language(DDL)* commands are parsed by a *DDL processor* and passed to the *execution engine*, which then goes through the *index/file/record manager* to alter the *metadata*, that is[13], the schema information for the database.

Overview of Query Processing

The great majority of interactions with the DBMS follow the path on the left side of Figure 8.1. A user or an application program initiates some action[14], using the *data-manipulation language(DML)*. This command does not affect the schema of the database, but may affect the content of the database(if the action is a modification command) or will extract data from the database(if the action is a query). DML statements are handled by two separate subsystems: answering the query and transaction processing.

trail[treil]*n*.（长串的）踪迹;痕迹

registrar['redʒɪˌstrɑ:] *n*.（大学的）教务长,教务主任

profound[prə'faund] *adj*.巨大的;深远的

parse[pɑ:s]*v.*（对句子）作语法分析;作句法分析

extract[iks'trækt]*v.* 提取;抽取

---

[9]　Ask for sth:to say that you want sb to give you sth;要求某事物。

[10]　同:...a person or persons who is/are responsible for... .

[11]　同:...its trail that begins at the upper right side of ...

[12]　Since(conj.):because;as;因为；由于。

[13]　That is:used to say what sth means or to give more information;也就是说；即；换句话说。

[14]　Action[U]:the process of doing sth in order to make sth happen or to deal with a situation;行为；行为过程。

Storage and Buffer Management

The data of a database normally resides in *secondary storage*; in today's computer systems "secondary storage" generally means magnetic disk. However, to perform any useful operation on data, that data must be in *main memory*. It is the job of the *storage manager* to control the placement of data on disk and its movement between disk and main memory.

In a simple database system, the storage manger might be nothing more than the file system of the underlying operating system. However, for efficiency purposes, DBMS's normally control storage on the disk directly, at least under some circumstances. The storage manger keeps track of the location of files on the disk and obtains the block or blocks containing a file on request from the buffer manager.

The *buffer manager* is responsible for partitioning the available main memory into *buffers*, which are *page-sized regions* into which *disk blocks* can be transferred. Thus, all DBMS components that need information from the disk will interact with the buffers and the buffer manger, either directly or through the execution engine. The kinds of information that various components may need include:*data*, *metadata*, *log records*, *statistics*, and *indexes*.

Transaction Processing

It[15] is normal to group one or more database operations into a *transaction*, which[16] is a unit of work that must be executed atomically and in apparent isolation from other transactions. In addition, a DBMS offers the guarantee of durability:that the work of a completed transaction will never be lost. The *transaction manager* therefore accepts *transaction commands* from an application, which tell the transaction manager when transactions begin and end, as well as information about the expectations of the application. The transaction processor performs the following tasks: *logging*, *concurrency control*, and *deadlock resolution*.

underlying[ˌʌndəˈlaiiŋ]*adj.*底层的

partition[pɑːˈtiʃən]*v.* 分割,隔开

statistic[stəˈtistik]*n.* 统计数据

atomically[əˈtɒmɪkəlɪ] *adv.*整体地;不可再分地
isolation[ˌaɪsəˈleɪʃən] *n.*隔离;隔离状态
guarantee[ˌgærənˈtiː] *n.*保障

---

[15] It:used in the position of the subject or object of a verb when the real subject or object is at the end of sentence;用作形式主语或形式宾语，而真正的主语或宾语在句末。

[16] Tip:Here the relative pronoun "which" refers to "a transaction", and the subject of the relative clause is also "a transaction". Note that we must include a relative pronoun in a non-defining relative clause.

The Query Processor

The portion of the DBMS that most affects the performance that the user sees is the query processor. In Figure 8.1 the query processor is represented by two component:

1. The *query compiler*, which translates the query into an internal form called a *query plan*. The latter is a sequence of operations to be performed on the data. Often the operations in a query plan are implementations of "relational algebra" operations. The query compiler consists of three major units:

    a. A *query parser*, which builds a tree structure from the textual form of the query.

    b. A *query preprocessor*, which performs *semantic checks* on the query, and performing some tree transformations to turn the parse tree into a tree of algebraic operators representing the initial query plan.

    c. A *query optimizer*, which transforms the initial query plan into the best available sequence of operations on the actual data.

2. The *execution engine*, which has the responsibility for executing each of the steps in the chosen query plan. The execution engine interacts with most of the other components of the DBMS, either directly or through the buffers. It most get the data from the database into buffers in order to manipulate that data. It needs to interact with the *scheduler* to avoid [17] accessing database that is locked, and with the log manager to make sure that all database changes are properly logged.

algebra['ældʒibrə]*n.*
代数

TERMINOLOGY

atomicity   main memory

---

[17] Avoid—[V -ing]。

buffer
buffer manager
column
concurrency control
control
data
data flow
data structure
database
database administrator(DBA)
database management system(DBMS)
database system
data-definition language(DDL)
data-manipulation language(DML)
DDL processor
deadlock resolution
disk block
durability
execution engine
index/file/record manager
index
isolation
log record
logging

metadata
modify
page-sized region
query
query compiler
query language
query optimizer
query parser
query plan
query preprocessor
query processor
relation
scheduler
schema
schema-altering command
secondary storage
semantic check
statistic
storage manager
system component
table
transaction
transaction command
transaction manager

EXERCISES

8.1 Translate each of the following key terms:

a)table
b)database
c)query language
d)semantic check
e)database management system
f)logging
g)metadata
h)column

8.2 Fill in the blanks in each of the following statements:

a)A(n) _____ is a collection of related data.

b)Logical structures of data are also called _____.

c)A data manipulation language is also known as a(n) _____.

d)A(n) _____ is a program that sets up, or structures, a database.

e)The transaction processor performs the tasks:_____, concurrency control, and deadlock resolution.

f)Data-definition commands are parsed by a _____.

g)The kinds of information that all DBMS components may need include data, metadata, log records, statistics, and _____.

h)A(n) _____ refers to a person responsible for structuring, coordinating, linking, and maintaining databases.

i)Query languages are also known as _____.

j)In a DBMS, the query processor consists of a _____ and a execution engine.

k)The most prominent data manipulation language today is _____.

8.3 State whether each of the following is *true* or *false*. If *false*, explain why.

a)A data management system acts as the interface between application programs and data.

b)DML commands can affect the schema of a database.

c)Buffers are page-sized regions into which disk blocks can be transferred.

d)The data of a database normally resides in primary storage.

e)DML statements are handled by two separate subsystems:answering the query and transaction processing.

f)A transaction is a unit of work that must be executed atomically and in apparent isolation from other transactions.

g)A query compiler consists of two major units:a query parser and a query optimizer.

8.4 Match each numbered item with the most closely related lettered item:

a)database	1.recovery of a database in the face of failures, errors of many kinds, or intentional misuse.
b)DBMS	2.component of a database management system that defines each data element as it appears in the database.
c)data definition language	3.a field.

d)data manipulation language	4.standard data manipulation language for relational database management systems.
e)query	5.language associated with a database management system that is employed by end users and programmers to manipulate data in the database.
f)durability	6.question about data.
g)SQL	7.special software to create and maintain a database and enable individual business applications to extract the data they need without having to create separate files or data definitions in their computer programs.
k)data element	8.collection of data organized to service many applications at the same time by storing and managing data so that they appear to be in one location.

8.5 Expand each of the following acronyms:

a)DBMS.

b)DDL.

c)DBA.

d)DML.

Reading Material （阅读材料）

Social Impact of Database Technology

In the past, collections of data were processed manually, meaning that relationships between information scatted throughout large collections were essentially undiscoverable. For example, the reading habits of a library's patrons might have been buried within the library's records, but unraveling it would have been a time-consuming process. Today, however, most library records are automated, and in some cases, profiles of an individual's reading habits are within easy reach. It is now feasible for libraries to provide such information to marketing firms, law enforcement agencies, political parties, employers, and private individuals.

This example is representative of the potential problems that permeate the entire spectrum of database applications. Technology has made it easy to collect enormous amounts of data and to merge or compare different data collections to obtain relationships that would otherwise remain buried in the heap. The ramifications, both positive and negative, are enormous.

These ramifications are not merely a subject of academic debate—they are realities. Data collection is now conducted on a massive scale. In some cases the process is readily apparent; in others it is subtle. Examples of the first case occur when one is explicitly asked to provide information. This might be done in a voluntary manner, as in surveys or contest registration forms, or it might be done in an involuntary manner, such as when imposed by government regulations. Sometimes whether it is voluntary or not depends on one's point of view. Is providing personal information when applying for a loan voluntary or involuntary? The distinction depends on whether receiving the loan is a convenience or a necessity. To use a credit card at some retailers now requires that you allow your signature to be recorded in a digitized format. Again, providing the information is either voluntary or involuntary depending on your situation.

More subtle cases of data collection avoid direct communication with the subject. Examples include a credit company that records the purchasing practices of the holders of its credit cards, sites on the World Wide Web that record the identities of those who visit the site, and social activists who record the license plate numbers on the cars parked in a targeted institution's parking lot. In these cases the subject of the data collection might not be aware that information is being collected and less likely to be aware of the existence of the databases being constructed.

Sometimes the underlying data-collection activities are self-evident if one merely stops to link. For example, a grocery store might offer discounts to its regular customers who register in advance with the store. The registration process might involve the issuance of identification cards that must be presented at the time of purchase to obtain the discount. The result is that the store is able to compile a record of the customers' purchases—a record whose value far exceeds the value of the discounts awarded.

Of course, the force driving this boom in data collection is the value of the data, which is amplified by advances in database technology that allow data to be linked in ways that reveal information that would otherwise remain obscure. For example, the purchasing patterns of credit card holders can be classified and cross-listed to obtain customer profiles of immense marketing value. Subscription forms for body-building magazines can be mailed to those who have recently purchased exercise equipment, whereas subscription forms for dog obedience magazines can be targeted toward those who have recently purchased dog food. Alternative ways of combining information are sometimes very imaginative. Welfare records have been compared to criminal records to find and apprehend parole violators, and in 1984 the Selective Service in the United States used old birthday registration lists from a popular ice cream retailer to identify citizens who had failed to register for the military draft.

There are several approaches that can be taken to protect society from abusive use of databases. One is to apply legal remedies. Unfortunately, passing a law against an action does not stop the action from occurring but merely makes the action illegal. A prime example in the United States is the Privacy Act of 1974 whose purpose was to protect citizens from abusive use of government databases. One provision of this act required government agencies to publish notice of their databases in the Federal Register to allow citizens to access and correct their personal information. However, government agencies have been slow to comply with this provision. This does not necessarily imply malicious intent. In many cases the problem has been one of bureaucracy. But, the fact that a bureaucracy might be constructing personnel databases that it is unable to identify is not reassuring.

Another, and perhaps more powerful, approach to controlling database abuse is public opinion. Databases will not be abused if the penalties outweigh the benefits; and the penalty businesses fear the most is adverse public opinion—this goes right to the bottom line. In the early 1990s it was public opinion that ultimately stopped major credit bureaus from selling mailing lists for marketing purposes. More recently, America Online(a major Internet service provider) buckled under public pressure against its policy of selling customer-related information to telemarketers. Even government agencies have bowed to public opinion. In 1997 the Social Security Administration in the United States modified its plan to make social security of the information. In these cases results were obtained in days—a stark contrast to the extended time periods associated with legal processes.

Of course, in many cases databases applications are beneficial to both the holder and the subject of the data, but in all cases there is a loss of privacy that should not be taken lightly. Such privacy issues are serious when the information is accurate, but they become gigantic when the information is erroneous. Imagine the feeling of hopelessness if you realized that your credit rating was adversely affected by erroneous information. Imagine how your problems would be amplified in an environment in which this misinformation was readily shared with other institutions.

Privacy problems are, and will be, a major side effect of advancing technology in general and database techniques in particular. The solutions to these problems will require an educated, alter, and active citizenry.

Note

Lesson 9 Basics of the Relational Model

Dimensional（形容词）
构成…维的

The *relational model* gives[1] us a single way to represent data:as a two-dimensional *table* called a *relation*. Figure 9.1 is an example of a relation, which we shall call Movies. The rows each[2] represent a movie, and the columns each represent a property of movies. In this section, we shall introduce the most important terminology regarding[3] relations, and illustrate them with the Movies relation.

illustrate['iləstreit]*v.*
（用示例、图表等）说
明;解释

title	year	length	genre
Gone With the Wind	1939	231	drama
Star Wars	1977	124	sciFi
Wayne's World	1992	95	comedy

Figure 9.1   The relation Movies

Attributes

The *columns* of a relation are named by *attributes*; in Figure 9.1 the attributes are title, year, length, and genre. Attributes appear at the tops of the columns. Usually, an attribute describes the meaning of *entries* in the column below. For instance, the column with attribute length holds the length, in minutes, of each movie.

Schemas

The name of a relation and the set of attributes for a relation is called the *schema* for that relation. We show the schema for the relation with the relation name followed by a parenthesized list of[4] its attributes. Thus, the schema for relation Movies of Figure 9.1 is

parenthesized[pə're
nθisaizd]*adj.*补充的;插
入说明的

Movies(title, year, length, genre).

The attributes in a relation schema are a *set*, not a *list*. However, in order to talk about[5] relations we often must specify a "standard"

---

[1] Give—[VNN]。

[2] Tip:when "each" (det.) is used after a plural subject, it has a plural verb;each 用于复数主语后，谓语动词用复数。

[3] Regarding(prep.):concerning sb/sth; about sb/sth;关于；至于。

[4] Of:used to say what sth is, consists of, or contains; （用于表示性质、组成或涵盖）即，由…组成。

[5] About(prep.):on the subject of sb/sth; in connection with sb/sth;关于；对于。

order for the attributes, as above[6], we shall take this ordering to be the standard order whenever we display the relation or any of its rows.

In the relational model, a *database* consists of one or more relations. The set of schemas for the relations of a database is called a *relational database schema*, or just a *database schema*.

Tuples

The *rows* of a relation, other than[7] the header row containing the attribute names, are called *tuples*. A tuple has one component for each attribute of the relation. For instance, the first of the three tuples in Figure 9.1 has the four components Gone With the Wind, 1939, 231, and drama for attributes title, year, length, and genre, respectively[8]. When we wish to write a tuple in isolation, not as part of a relation, we normally use commas to separate components, and we use parentheses[9] to surround the tuple. For example,

(Gone With the Wind, 1939, 231, drama)

is the first tuple of Figure 9.1. Notice that when a tuple appears in isolation, the attributes do not appear, so some indication of the relation to which[10] the tuple belongs must be given. We shall always use the order in which[11] the attributes were listed in the relation schema.

Domains

The relational model requires that each component of each tuple be atomic; that is, it must be of some elementary type such as *integer* or *string*. It[12] is not permitted for a value to be a record structure, set, list, array, or any other type that reasonably can have its values broken into smaller components.

It is further assumed that associated with each attribute of a relation is a *domain*, that is, a particular elementary type. The components of any tuple of the relation must have, in each

comma['kɔmə]*n.*逗号
surround[sə'raund]*v.*
环绕;围绕

---

[6] Above(adv.):at or to a higher place;在（或向）上面；在（或向）较高处。

[7] Other than:except;除…以外。

[8] Respectively(adv.):in the same order as the people or things already mentioned;分别；依次为。

[9] Parenthesis:a word, sentence, etc. that is added to a speech or piece of writing, especially in order to give extra information. In writing, it is separated from the rest of the text using brackets, commas or DASHES;插入语。

[10] OR:*(less formally)*... some indication of the relation which the tuple belongs to must be...

[11] Tip:Here the relative pronoun "which" refers to "the order", and the subject of the relative clause is "the attributes".

[12] It:used in the position of the subject or object of a verb when the real subject or object is at the end of sentence;用作形式主语或形式宾语，而真正的主语或宾语在句末。

component, a value that belongs to the domain of the corresponding column. For example, tuples of the Movies relation of Figure 9.1 must have a first component that is a string, second and third components that are integers, and a fourth component whose value is a string.

It is possible to include the domain, or *data type*, for each attribute in a relation schema. We shall do so by appending a colon[ˈkəulən]*n.*冒号   colon and a type after attributes. For example, we could represent the schema for the Movies relation as:

Movies(title:string, year:integer, length:integer, genre:string)

Equivalent Representations of a Relation

Relations are sets of tuples, not lists of tuples. Thus the order in which the tuples of a relation are presented is immaterial[ˌiməˈtɪəri: əl]*adj.*无关紧要的;不相 干的   immaterial. For example, we can list the three tuples of Figure 9.1 in any of their six possible orders, and the relation is "the same" as Figure 9.1.

Moreover, we can reorder the attributes of the relation as we choose, without changing the relation. However, when we reorder the relation schema, we must be careful to remember that the attributes are *column headers*. Thus, when we change the order of the attributes, we also change the order of their columns. When the columns move, the components of tuples change their order as well. The result is that each tuple has its components permuted in the same way as the attributes are permuted.

permute[pə(:)ˈmjuːt] *v.*改变序列,组合;置换   For example, Figure 9.2 shows one of the many relations that could be obtained from Figure 9.1 by permuting rows and columns. These two relations are considered[13] "the same". More precisely, these two tables are different presentations of the same relation.

year	genre	title	length
1977	sciFi	Star Wars	124
1992	comedy	Wayne's World	95
1939	drama	Gone With the Wind	231

Figure 9.2 Another presentation of the relation Movies

[13] Consider—[VN -N]。

Relation Instances

A relation about movies is not static; rather[14], relations change over time. We expect to insert tuples for new movies, as these appear. We also expect changes to existing tuples if we get revised or corrected information about a movie, and perhaps deletion of[15] tuples for movies that are expelled from the database for some[16] reason.

It is less common for the schema of a relation to change. However, there are situations where we might want to add or delete attributes. Schema changes, while possible in commercial database systems, can be very expensive, because each of perhaps millions of tuples needs to be rewritten to add or delete components. Also, if we add an attribute, it may be difficult or even impossible to generate appropriate values for the new components in the existing tuples.

We shall call a set of tuples for a given relation an *instance* of that relation. For example, the three tuples shown in Figure 9.1 form an instance of relation Movies. Presumably, the relation Movies has changed over time and will continue to change over time. For instance, in 1990, Movies did not contain the tuple for Wayne's World. However, a conventional database system maintains only one version of any relation:the set of tuples that are in the relation "now". This instance of the relation is called the *current instance*.

Keys of Relations

There are many constraints on relations that the relational model allows us to place on database schemas. However, one kind of constraint is so fundamental that[17] we shall introduce it here:*key* constraints. A set of attributes forms a key for a relation if we do not allow two tuples in a relation instance to have the same values in all the attributes of the key.

static['stætik]*adj.* 静 态的;不变化的
existing[ig'zistiŋ]*adj.* 现存的;存在的

expel[iks'pel]*v.* 驱 逐 ; 赶走

---

[14] Rather(adv.):used to introduce an idea that is different or opposite to the idea that you have stated previously;（提出不同或相反的观点）相反，反而，而是。

[15] Of:used after nouns formed from verbs. The noun after "of" can be either the object or the subject of the action;用于由动词转化的名词之后，of 之后的名词可以是受动者，也可以是施动者。

[16] Some(det.):used with singular nouns to refer to a person, place, thing or time that is not known or not identified;（与单数名词连用，表示未知或未确指的人、地、事物或时间）某个。

[17] So...that...:in such a way that...;这样…为的是…；如此…以致…。

declare[diˈklɛə]*v.*
表明;宣传;断言

Example 9.1: We can declare that the relation Movies has a key consisting of the two attributes title and year. That is, we don't believe there could ever be two movies that had both the same title and the same year. Notice that title by itself does not form a key, since sometimes "remarkes" of a movie appear. For example, there are three movies named King Kong, each made in a different year. It should also be obvious that year by itself is not a key, since there are usually many movies made in the same year.

indicate[ˈindikeit]*v.*
标示;显示（信息）

We indicate the attribute or attributes that form a key for a relation by underlining the key attribute(s). For example, the Movies relation could have its schema written as:

Movies(title, year, length, genre).

Remember[18] that the statement that a set of attributes forms a key for a relation is a statement about all possible instances of the relation, not a statement about a single instance. For example, looking only at the tiny relation of Figure 9.1, we might imagine that genre by itself forms a key, since we do not see two tuples that agree on the value of their genre components. However, we can easily imagine that if the relation instance contained more movies, there would be many dramas, many comedies[19], and so on. Thus, there would be distinct tuples that agreed on the genre component. As a consequence, it would be incorrect to assert that genre is a key for the relation Movies.

imagine[iˈmædʒin]*v.*
想象;设想

assert[əˈsəːt]*v.* 明确肯定;断言

While we might be sure that title and year can serve as a key for Movies, many real-world databases use *artificial keys*, doubting that it is safe to make any assumption about the values of attributes outside[20] their control. For example, companies generally assign employee ID's to all employees, and these ID's are carefully chosen to be unique numbers. One purpose of these ID's is to make sure[21] that in the company database each employee can be distinguished form all others, even if[22] there are several employees with the same name.

---

[18] Remember—[V (that)]。
[19] Drama:戏剧;comedy:喜剧。
[20] Outside(prep.):not part of sth;不在…范围内；不属于。
[21] Make sure(adj.) that/of sth:to do sth in order to be certain that sth happens;确保；设法保证。
[22] Even if/though:despite the fact or belief that;no matter whether;即使；纵然。

TERMINOLOGY

artificial key	key
attribute	list
column	relation
column header	relational database schema
current instance	relational model
data type	row
database	schema
database schema	set
domain	string
entry	table
instance	tuple
integer	

EXERCISES

9.1 Translate each of the following key terms:

    a)relational model

    b)row

    c)column

    d)table

    e)tuple

    f)attribute

    g)data type

    h)key

    i)domain

    j)schema

9.2 Fill in the blanks in each of the following statements:

    a)The _____ database is the most widely used database structure.

    b)The columns of a relation are named _____.

    c)A _____ consists of relations.

d)The name of a relation and the set of attributes for a relation is called _____.

e)The relational model represents data as a two-dimensional table called a _____.

f)The set of relation schemas of a database is called a _____.

g)The rows of a relation, other than the header row are called _____.

h)A set of attributes forms a _____ that uniquely identifies each tuple.

9.3 State whether each of the following is *true* or *false*. If *false*, explain why.

a)The relational model requires that each component of each tuple be atomic.

b)The attributes in a relation schema are a list, not a set.

c)In the relational model, a database consists of one or more relations.

d)A domain refers to a particular data type.

e)The order in which the tuples of a relation are presented is material.

9.4 Match each numbered item with the most closely related lettered item:

a)tuple	1.a set of tuples for a given relation.
b)attributes	2.a particular elementary type.
c)domain	3.a set of attributes forms uniquely identifies each tuple.
d)relation instance	4.has one component for each attribute of the relation.
e)key	5.appear at the tops of the columns and describe the meaning of entries in the column below.

9.5 There are instances of two relations that might constitute part of a banking database. Indicate the following:

The relation Accounts

acctNo	type	balance
12345	savings	12000
23456	checking	1000
34567	savings	25

The relation Customers

firstName	lastName	idNo	account
Robbie	Banks	901-222	12345
Lena	Hand	805-333	12345
Lena	Hand	805-333	23456

a)The attributes of each relation.

b)The tuples of each relation.

c)The components of one tuple from each relation.

d)The relation schema for each relation.

e)The database schema.

f)A suitable domain for each attribute.

g)Another equivalent way to present each relation.

Reading Material（阅读材料）

SQL

Structured Query Language(SQL) is the standard language for relational systems, and it is supported by just about every database product on the market today. SQL was originally developed by IBM Research in the early 1970s; it was first implemented on a large scale in an IBM prototype called System R, and subsequently reimplemented in numerous commercial products from both IBM and many other vendors. Here we present an overview of the major features of the SQL language. Throughout our discussion, we take the unqualified name SQL to refer to the current version of the standard, barring explicit statements to the contrary.

```
CREATE TYPE S# ... ;
CREATE TYPE NAME ...;
CREATE TYPE P# ... ;
CREATE TYPE COLPR ... ;
CREATE TYPE WEIGHT ... ;
CREATE TYPE QTY ... ;

CREATE TABLE S
    { S#          S#;
      SNAME       NAME,
      STATUS      INTEGER,
      CITY        CHAR(15),
       PRIMARY KEY (S#) };

CREATE TABLE P
    {  P#          P#;
      PNAME       NAME,
      COLOR       COLOR;
      STATUS      INTEGER,
      CITY         CHAR(15),
       PRIMARY KEY (P#) };

CREATE TABLE SP
    { S#          S#;
      P#          P#,
      QTY         QTY,
       PRIMARY KEY (S#,P#)'
       FOREIGN KEY (S#) REFERENCES S,
       FOREIGN KEY (P#) REFERENCES P}
```

Figure 9.3   The suppliers-and-parts database(SQL definition)

SQL includes both data definition and data manipulation operations. We consider *definitional operations* first. Figure 9.3 gives an SQL definition for the suppliers-and-parts database. As you can see, the definition includes one *CREATE TYPE statement* for each of the six *user-defined types(UDTs)* and one *CREATE TABLE statement* for each of the three base tables. Each such CREATE TABLE statement specifies the name of the base table to be created, the names and types of the columns of that table, and the *primary key* and any *foreign keys* in that table. Also, please note the following:

1.   We often make use of the "#" character in *type names* and *column names*, but in fact that character is not legal in the standard.

2.   We use the semicolon ";" as a *statement terminator*. Whether SQL actually uses such terminators depends on the context. The specifics are beyond the scope of this section.

3.   The built-in type CHAR in SQL requires an associated length—15 in the figure—to be specifies.

Having defined the database, we can now start operating on it by means of the SQL *manipulative operations* SELECT, INSERT, DELETE, and UPDATE. In particular, we can perform relational *restrict*, *project*, and *join operations* on the data, in each case by using the SQL data manipulation statement SELECT. Some examples are given in Figure 9.4.

Restrict:				Result:	

Restrict: Result:

SELECT   S#, P#, QTY

FROM   SP

WHERE   QTY<QTY(150);

S#	P#	QTY
S1	P5	100
S1	P6	100

Project: Result:

SELECT   S#, CITY

FROM   S;

S#	CITY
S1	London
S1	Paris
S3	Paris
S4	London
S5	Athens

Join:

SELECT   S.S#, SNAME, STATUS, CITY, P#, QTY

FROM   S, SP

WHERE   S.S# = SP.S#;

Result:

S#	SNAME	STATUS	CITY	P#	QTY
S1	Smith	20	London	P1	300
S1	Smith	20	London	P2	200
S3	Smith	20	London	P3	400
...	...	...	...	...	...
S5	Clark	20	London	P5	400

Figure 9.4   Restrict, project, and join examples in SQL

We remark that SQL also supports a shorthand form of the SELECT clause as illustrated by the following example:

SELECT * FROM S;   /* or "SELECT S.*" (i.e., the "*" can be dot-qualified) */

The result is a copy of the entire S table; the *star* or *asterisk* is shorthand for a "commalist" of (a) names of all columns in the first table referenced in the FROM clause, in the left-to-right order in which those columns are defined within that table, followed by (b) names of all columns in the second table referenced in the FROM clause, in the left-to-right order in which those columns are defined within that table(and so on for all of the other tables referenced in the FROM clause). Note:The expression SELECT * FROM T, where T is a table name, can be further abbreviated to just TABLE T.

Turning now to *update operations*. Like SELECT, however, INSERT, DELETE, and UPDATE are all *set-level operations*, in general. Here are some set-level update examples for the suppliers-and-parts database:

```
INSERT
INTO    TEMP      ( P# , WEIGHT)
        SELECT    P# , WEIGHT
        FROM      P
        WHERE     COLOR = COLPR ('Red') ;
```

This example assumes that we have already created another table TEMP with two columns, P# and WEIGHT. The INSERT statement inserts into that table part numbers and corresponding weights for all red parts.

```
DELETE
FROM      SP
WHERE     P# = P# ('P2') ;
```

This DELETE statement deletes all shipments for part P2.

```
UPDATE    S
SET       STATUS = 2 * STATUS '
          CITY = 'Rome'
WHERE     CITY = 'Paris' ;
```

This UPDATE statement doubles the status for all suppliers in Paris and moves those suppliers to Rome.

Note:SQL does not include a direct analog of the relational assignment operation. However, we can simulate that operation by first deleting all rows from the target table and then performing an INSERT ... SELECT ...(as in the first example above)into that table.

Note

Lesson 10 Network Architecture

Communications channels can be connected in different arrangements, or *networks*, to suit different users' needs[1]. A *computer network* is a communications system connecting[2] two or more computers that work together to exchange information and share resources. *Network architecture* describes how the network is arranged and how the resources are coordinated and shared.

channel['tʃænl]*n.* 途径;渠道;系统

coordinate[kəu'ɔ:din eit]*v.*协调;使…相配合

There are a number of[3] specialized terms that describe computer networks. Some terms often used with network are:

- **Node**—any device that is connected to a network. It could be a computer, printer, or data storage device.
- **Client**—a node that requests and uses resources available from other nodes[4]. Typically, a client is a user's *microcomputer*.
- **Server**—a node that shares resources with other nodes. Depending on[5] the resources shared, it may be called a *file server*, *printer server*, *communication server*, *Web server*, or *database server*.
- **Network operating systems(NOS)**—controls and coordinates the activities of all computers and other devices on a network. These activities include electronic communication and the sharing of[6] information and resources.

sharing['ʃεəriŋ]*n.*共享

- **Distributed processing**—a system in which[7] computing power is located and shared at different locations. This type of system is common in *decentralized organizations* where[8] divisional offices have their own computer

---

[1] Need[C, usually pl.]:the things that sb requires in order to achieve what they want;想要的事物。

[2] 同:...a communications system that connects two or more computers...

[3] A number of=some.

[4] Tip:... resources available from other nodes.

[5] Depending on:according to;决定于; 视乎。

[6] Of:used after nouns formed from verbs. The noun after 'of' can be either the object or the subject of the action;用于由动词转化的名词之后，of之后的名词可以是受动者，也可以是施动者。

[7] Tip:Here the relative pronoun "which" refers to "a system", and the subject of the relative clause is "computing power".

[8] OR:*(more formally)*...decentralized organizations in which divisional offices...

network['netwə:k]v.
（计算机）将…连接成
网络

integrate['intigreit]v.
使合并;成为一体

self-contained[ˌselfk
ən'teind]adj.自治的
dispersed[dis'pə:st]
adj.分治的
geographical[dʒiə'gr
æfikəl]adj.地理上的

peripheral[pə'rifərəl]
adj.外围的;非核心的

handle['hændl]v.处理;
应对

systems. The computer systems in the divisional offices are networked to the organization's *main* or *centralized computer*.

● *Host computer*—a large centralized computer, usually a *minicomputer* or a *mainframe*.

A network may consist only of microcomputers, or it may integrate microcomputers or other devices with larger computers. Networks can be controlled by all nodes working together equally or by specialized nodes coordinating and supplying all resources. Networks may be simple or complex, self-contained or dispersed over a large geographical area.

A network can be arranged or configured in several different ways. This arrangement is called[9] the network's *topology*. The four principal network topologies are star, bus, ring, and hierarchical.

In a *star network*, a number of small computers or peripheral devices are linked to a central unit. This central unit may be a *host computer* or a *file server*.

All communications pass through this central unit. Control is maintained by *polling*. That is[10], each connecting device is asked("polled") whether it has a message to send. Each device is then in turn[11] allowed to send its message.

One particular advantage of the star form of network is that it can be used to provide a *time-sharing system*. That is, several users can share resources("time") on a central computer. The star is a common arrangement for linking several microcomputers to a mainframe that allows access to an organization's database.

In a *bus network*, each device in the network handles its own communications control. There is no host computer. All communications travel along a common connecting cable called a *bus*. As the information passes along the bus, it[12] is examined by each device to see if the information is intended[13] for it.

---

[9] Call—[VN -N]。

[10] That is:used to say what sth means or to give more information;也就是说；即；换句话说。

[11] In turn:one after the other in a particular order;依次；轮流；逐个。

[12] It:used in the position of the subject or object of a verb when the real subject or object is at the end of sentence;用作形式主语或形式宾语，而真正的主语或宾语在句末。

[13] Intended(adj.) for sb/sth|~ as sth|~ to be/do sth:planned or designed for sb/sth;为…打算（或设计）的。

The bus network is typically used when only a few microcomputers are to be linked together. This arrangement is common for sharing data stored on different microcomputers. The bus network is not as efficient as[14] the star network for sharing common resources. (This is because the bus network is not a direct link to the resources.) However, a bus network is less expensive and is in very common use.

In a *ring network*, each device is connected to two other devices, forming[15] a ring. There is no central file server or computer. Messages are passed around the ring until they reach the correct destination. With microcomputers, the ring arrangement is the least frequently used of[16] the four geographical areas. These mainframes tend to operate fairly autonomously. They perform most or all of[17] their own processing and only occasionally share data and programs with other mainframes.

A ring network is useful in a decentralized organization because it makes[18] possible a *distributed data processing system*. That is, computers can perform processing tasks at their own dispersed locations. However, they can also share programs, data, and other resources with each other.

The *hierarchical network*—also called a *hybrid network*—consists of several computers linked to a central host computer, just like a star network. However, these other computers are also hosts to other, smaller computers or to peripheral devices.

Thus, the host at the top of the hierarchy could be a mainframe. The computers below[19] the mainframe could be minicomputers, and those below, microcomputers[20]. The hierarchical network allows various computers to share databases, processing power, and different output devices.

destination[ˌdestiˈnei∫ən]*n.*目的地

autonomously[ɔːˈtɒnəməs]*adv.*自治地;自主地

occasionally[əˈkeiʒənəli]*adv.*偶尔;间或

---

[14] As(adv.)… as…(prep./conj.):used when you are comparing two people or things, or two situations;（比较时用）像…一样，如同。
[15] 同:…, so the devices forms a ring.
[16] Of:belonging to sth; being part of sth; relating to sth;属于（某物）；关于（某物）。
[17] Of:used to show sb/sth belongs to a group, often after *some, a few*,etc.;（常用在 some、a few 等词语之后，表示人或物的所属）属于…的。
[18] Make—[VN -ADJ]。
[19] Below(prep.):of a lower rank or of less important than sb/sth;（级别、重要性）低于。
[20] 同:…, and those below the minicomputers could be microcomputers.

A hierarchical network is useful in centralized organizations. For example, different departments within an organization may have individual microcomputers connected to departmental minicomputers. The miniconputers in turn may be connected to the corporation's mainframe, which contains data and programs accessible to all.

For a summary of the network configurations, see Table 10.1.

Table 10.1 Principal network configurations

Type	Description
Star	Several computers connected to a central server or host; all communications travel through central server;good for sharing common resources
Bus	Computers connected by a common line; communication travels along this common line; less expensive than star
Ring	Each computer connected to two others forming a ring; communications travel around ring; often used to link mainframe computers in decentralized organizations
Hierarchical	One top-level computer connected to next-level computers, which are connected to third-level computers; often used in centralized organizations

Every network has a *strategy*, or way of coordinating the sharing of information and resources. The most common network strategies are terminal, peer-to-peer, and client/serer systems.

In a *terminal network system*, processing power is centralized in one large computer, usually a mainframe. The nodes connected to this host computer are either *terminals*, with little or no processing capabilities, or microcomputers running special software that allows them to act as terminals. The star and hierarchical networks are typical configurations with UNIX as the operating system.

Many airlines reservation systems are terminal systems. A large central computer maintains all the airline schedules, rates, seat availability, and so on. Travel agents use terminals to schedule reservations. Although the tickets may be printed along with travel itineraries at the agent's desk, nearly all processing is done at the central computer.

itinerary[aiˈtinəˌreri:]
n.行程;旅行日程

One advantage of terminal network systems is the centralized location and control of [21] technical personnel, software, and data. One disadvantage is the lack of control and flexibility for the end user. Another disadvantage is that terminal systems do not use the full processing power available with microcomputers. Though the terminal strategy was once very popular, most new systems do not use it.

In a *peer-to-peer network system*, nodes have equal *authority* and can act as both servers and clients. For example, one microcomputer can obtain files located on another microcomputer and can also provide files to other microcomputers. A typical configuration for a peer-to-peer system is the bus network. Commonly used network operating systems are Novell's NetWare Lite, Microsoft's Windows NT, and Apple's Macintosh Peer-to-Peer LANs.

There are several advantages to using this type of strategy. The networks are inexpensive and easy to install, and they usually work well for small systems with fewer than ten nodes. As[22] the number of nodes increase, however, the performance for the network declines. Another disadvantage is the lack of powerful management software to effectively monitor a large network's activities. For these reasons, peer-to-peer networks are typically used by small networks.

Client/server network systems use one computer to coordinate and supply services to other nodes on the network. The server provides access to resources such as Web pages, databases, application software, and hardware. This strategy is based on specialization. Server nodes coordinate and supply specialized services, and client nodes request the services. Commonly used network operating systems are Novell's NetWare, Microsoft's Windows NT, IBM's LAN Server, and Banyan Vines.

One advantage of client/server network systems is their ability to handle very large networks efficiently. This strategy is

---

[21] Of:used after nouns formed from verbs. The noun after "of" can be either the object or the subject of the action;用于由动词转化的名词之后，of 之后的名词可以是受动者，也可以是施动者。

[22] As(conj.):while sth else is happening;当···时；随着。

extensively used on the Internet. For example, when you connect to a Web site, you are the client and the Web site's computer is the server. Another advantage of client/server networks is the powerful network management software that monitors and controls the network's activities. The major disadvantages are the cost of installation and maintenance.

For a summary of the network strategies, see Table 10.2.

Table 10.2   Network strategies

Type	Description
Terminal	One large computer provides all processing, strong central control, limited flexibility, and control for users
Peer-to-peer	Computers act as both servers and clients, inexpensive and easy to install, works well in small networks
Client/server	Several clients or computers depend upon one server or computer to coordinate and supply services

evolve[i'vɔlv]v.逐步发展;逐步演变

Computer networks in organizations have evolved over time. Most large organizations have a wide range of different network configurations, operating systems, and strategies. These organizations are moving toward integrating or connecting all of these networks together. That way, a user on one network can access resources available throughout the company. This is called *enterprise computing*.

TERMINOLOGY

authority
bus
bus network
centralized computer
client
client/server network system
communication server
computer network
database server

microcomputer
minicomputer
network
network architecture
network operating system(NOS)
network's topology
node
peer-to-peer network system
polling

decentralized organization

distributed data processing system

distributed processing

enterprise computing

file server

hierarchical network

host computer

hybrid network

main computer

mainframe

printer server

ring network

server

star network

strategy

terminal

terminal network system

time-sharing system

Web server

EXERCISES

10.1 Translate each of the following key terms:

    a)network architecture

    b)client

    c)server

    d)host computer

    e)node

    f)star network

    g)network's topology

    h)bus network

10.2 Fill in the blanks in each of the following statements:

    a)_____ describes how the network is arranged and how the resources are coordinated and shared.

    b)A network can be arranged in several different ways called the network's _____.

    c)In a _____ network, a number of small computers or peripheral devices are linked to a central unit.

    d)A hierarchical network is also called a _____.

    e)The most common network strategies are terminal, peer-to-peer, and _____ system.

    f)In a _____ network system, nodes have equal authority and can act as both servers and clients.

    g)A(n) _____ is a communications system connecting two or more computers to exchange information and share resources.

h)_____ systems use one computer to coordinate and supply services to other nodes on the network.

i)In a bus network, all communications travel long a common connecting cable called a _____.

j)_____ means that a user on one network can access resources available throughout the company.

10.3 State whether each of the following is *true* or *false*. If *false*, explain why.

a)The bus arrangement is common for linking several microcomputers to a mainframe that allows access to an organization's database.

b)Peer-to-peer networks are typically used by small networks.

c)The terminal network strategy is extensively used on the Internet.

d)A ring network is useful in a decentralized organization, because it makes possible a distributed data processing system.

e)One advantage of client/server network system is their ability to handle very large networks efficiently.

f)A typical configuration for a peer-to-peer system is the star network.

g)The bus network is as efficient as the star network for sharing common resources.

h)A hierarchical network is useful in centralized organizations.

10.4 Match each numbered item with the most closely related lettered item:

a)node	1.controls and coordinates the activities of all computers and other devices on a network.
b)client	2.a large centralized computer.
c)server	3.any device that is connected to a network.
d)NOS	4.a system in which computing power is located and shared at different locations.
e)distributed processing	5.a node that requests and uses resources available from other nodes.
f)host computer	6.a node that shares resources with other nodes.

10.5 Expand each of the following acronyms:

a)NOS.

b)P2P.

c)LAN.

d)C/S.

e)WAN.

Reading Material （阅读材料）

Organizing a Web Site

The modern, interactive, full-color Web site is really just a milepost in a long list of computer interconnection technologies. It was not the first technology used to connect computers and it won't be the last.

In the earliest days of computing, circa 1955, computers were huge, single program devices that were used to crunch large numbers or manipulate large amounts of text data. The only network available in those days was the proverbial "SneakerNet", where a programmer or graduate student would output data to a tape drive, remove the reel and walk to the site of the second computer and mount the tape on a drive connect to this computer. It didn't take long for programmers and researchers to become dissatisfied with this approach.

In the early 1960s, the *Advanced Research Projects Agency(ARPA)* started a project with the purpose of finding a way to connect these *mainframe computers* so that data could be shared between them.

ARPA's first step was to create the hardware necessary to transfer electrical signals from one *end* of the computer to the other. The second step was to create the software that sits on each computer and controls the flow and interpretation of the data from the other machine. The result of this research effort was the *TCP/IP protocol*. In the early days TCP/IP was only one of several popular protocols. Because it was developed by a research project instead of by a company, there was no license fee for using it. That factor, combined with the fact that the fee-based technologies were not that much better, led to its near-universal adoption in the decades that followed.

All the technologies that we will describe run on top of TCP/IP. That is to say that the other programs contain *calls* to the TCP/IP *application-programming interface(API)* somewhere in the lower layers of their systems. The reason for this is that every computer that runs TCP/IP software can communicate with every other computer that runs this software.

TCP/IP is not really a protocol, but rather a suite consisting of dozens of protocols. It is named for two of the most important protocols, the *Transmission Control Protocol(TCP)* and the *Internet Protocol(IP)*. TCP/IP is composed of four layers:

- *Application layer*—This layer can be a simple *chat program* or a complete *online store* such as Amazon.com. It conducts its business by making calls to transport a layer's API.

- *Transport layer*—This layer sends chunks of data from one computer to another with guaranteed delivery by making calls to the internetwork layer's API.

- *Internetwork layer*—This layer breaks the chunks of data into small pieces called *datagrams* and sends them to the other computer by calling the network access layer's API.

- *Network access layer*—This layer controls the hardware that sends bits and bytes to other computers.

The layer were created to separate low-level functionality from higher-level functionality. This organization of functions enables you to remove a product that provides one layer and replace it with one that performs better, is easier to use, and so on. Figure 10.1 shows an illustration of this interaction between layers.

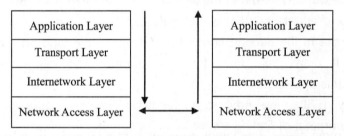

Figure 10.1   The various layers in the TCP/IP protocol work together to provide reliable intercomputer communications

Each layer calls the API of the layer beneath it to make requests for service. The end result is reliable communication between two computers.

The TCP and IP protocols take care of the movement of the data from one computer to the next. When the data gets there, however, the application layer must take over and process the data. A key part of that processing is the *HTTP protocol*.

A *protocol* is simple a published agreement between *clients* and *servers* that specifies what data will be passed from one party to the other and what syntax it will be in.

HTTP is an application layer protocol, which means that it depends on lower-layer protocols to do much of the work. In the case of HTTP, this lower-layer protocol is TCP/IP. The information in HTTP is transferred as *plain ASCII* characters. This is convenient for programmers because the instructions are human-readable, making debugging easier.

HTTP is a request and response protocol. This means that it is composed of pairs of *requests* and *responses*. Each of the *request-and-response pairs* is independent of every other pair. This kind of communication is called *connectionless*. If you want to string together requests to form a larger transaction, you must do this yourself in the programs that you write; HTTP will not do it fir you. Figure 10.2 shows where HTTP fits into the layer diagram that we drew in Figure 10.1.

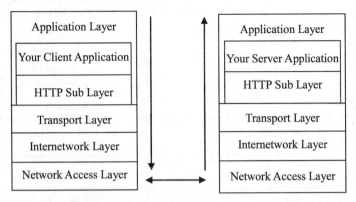

Figure 10.2   The HTTP protocol runs as a sublayer in the application layer

The format of the HTTP request is a *plain-text* message. The server reads the message and performs the task that it has been instructed to perform in the message. In some cases, this task is to retrieve a static document and send it back to the client's browser for display; but in other cases, this request is for a server-side application to be run. The results of this transaction, if any, are sent to the client.

On the left side of the Figure 10.2, you see a sublayer that is labeled "Your Client Application". In most cases, this client application is a browser. A *browser* is a piece of software that performs an amazing range of tasks. First, it is an HTTP application, meaning that it creates messages in that protocol for transmission to a Web server. In addition, it contains an *HTTP parser* that takes the messages that are returned by the Web server and translates them into a display that is pleasant, hopefully, to the human eye. This translation might require that a *.jpg image* be rendered, HTML be translated into text, *XML* be parsed and *Java Applets* be run, and so on.

The fact that the browser is normally a free *download*, or available as part of the operating system *installation pack* obscures the fact that it is really a sophisticated piece of software.

On the right side of Figure 10.2, you see a box labeled "Your Server Application". In the majority of installations, that server application is a piece of software known as the Web server. *Web servers* are also HTTP applications in that they both parse requests formatted in HTTP and respond by creating messages in HTTP format.

Originally, the job of the Web server was to locate documents and send them back to the requesting client computer in HTTP format. The Web servers were enhanced to add the capability to process CGI, Perl, and a host of other niche languages that composed the first generation of Web programming languages.

The modern Web server has a much more complicated task to perform. In addition to handing *static documents* to the client and running *scripts*, the Web server has to be able to call server-side applications, translate *dynamic Web pages*, and call *enterprise services*. Figure 10.3 shows a newer version of Figure 10.2, which shows the browser and the Web server in the application layer.

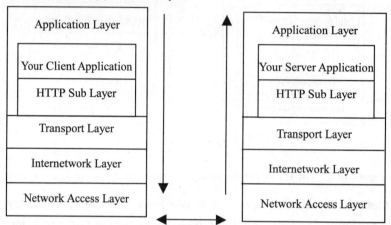

| Application Layer |
| Your Client Application |
| HTTP Sub Layer |
| Transport Layer |
| Internetwork Layer |
| Network Access Layer |

| Application Layer |
| Your Server Application |
| HTTP Sub Layer |
| Transport Layer |
| Internetwork Layer |
| Network Access Layer |

Figure 10.3   The Browser and the Web server are both HTTP applications

The client and the server applications are heavily dependent on the browser and the Web server's support mechanisms to run.

On the client side, the top sublayer in the application layer is unique in that it is almost always composed of an application that was downloaded from the Web server. These applications range from the simplest static HTML file to a complex dynamic Web page. In addition, the availability of script parsers in the browsers adds the capability of doing some computing outside of the Java-like applets as well.

In reality, the amount of applet processing that a browser performs is dwarfed by the amount of HTML, both static and dynamic, that it processes. The source of the static HTML is far from constant because every server-side, application-building programming language communicates the results to the browser that made the request using HTML and HTTP. Thus, form the server side, almost every Web site seems to send it HTML.

The top sublayer in the server side is far more complex than its counterpart on the client side. There is a seemingly endless variety of programming language that can be called by a Web server.

Note

Lesson 11 The Internet

internetwork[ˌintəˈnet
ˌwəːk]v.网络互联

The Internet is perhaps the most well-known, and the largest, implementation of[1] internetworking, linking hundreds of thousands of individual networks all[2] over the world. The Internet began as a U.S. Department of Defense[3] network to link scientists and university professors around the world. Even today individuals cannot connect directly to *the Net*, although anyone with a computer, a *modem*, and the willingness to pay a small monthly usage fee can access it through an Internet Service Provider. An *Internet Service Provider(ISP)* is a commercial organization with a permanent connection to the Internet through such popular *online services* as Prodigy and America Online[4] and through networks established by such giants as Microsoft and AT&T[5].

individual[ˌindiˈvidju
əl]n.个人

monthly[ˈmʌnθli]adj.
按月计算的

fee[fiː]n.费用

permanent[ˈpəːmənə
nt]adj.持续的;固定的

One of the most puzzling aspects of the Internet is that no one owns it and it has no formal management organization. As a creation of the Defense Department for sharing research data, this lack of[6] centralization was purposeful, to make it[7] less vulnerable to wartime or terrorist attacks. To join the Internet, an existing network needs only to pay a small registration fee and agree to certain[8] standards based on the *TCP/IP(Transmission Control Protocol/Internet Protocol)* reference model. Costs are low because the Internet owns nothing and so has no costs to offset. Each organization, of course, pays for its own networks and its own telephone bills, but those costs usually exist independent[9] of the Internet[10]. Regional Internet

purposeful[ˈpəːpəsfəl]
adj.有目的的;有意图的;
有意义的

vulnerable[ˈvʌlnərəbl]
adj.脆弱的;易受影响的

offset[ˈɔfset]v. 补偿;抵
消;耗费

---

[1] Of:used after nouns formed from verbs. The noun after "of" can be either the object or the subject of the action;用于由动词转化的名词之后，of之后的名词可以是受动者，也可以是施动者。

[2] All(adv.):completely;完全。

[3] The U.S. Department of Defense:美国国防部。

[4] America Online(AOL):美国在线。

[5] AT&T:美国电话电报公司，American Telephone & Telegraph 的缩写。

[6] Of:used after nouns formed from verbs. The noun after "of" can be either the object or the subject of the action;用于由动词转化的名词之后，of之后的名词可以是受动者，也可以是施动者。

[7] It:used in the position of the subject or object of a verb when the real subject or object is at the end of sentence;用作形式主语或形式宾语，而真正的主语或宾语在句末。

[8] Certain(adj.):used to mention a particular thing, person or group without giving any more details about it or them;（不提及细节时用）某事，某人，某种。

[9] Independent of sb/sth:not connected with or influenced by sth; not connected with each other;不相关的；不受影响的；无关联的。

[10] Tip:..., but those costs usually exist independent of the Internet.

companies have been established to which[11] member networks forward[12] all transmissions. These Internet companies route and forward all traffic, and the cost is still only that of a local telephone call. The result is that the costs of e-mail and other Internet connections tend to be far lower than equivalent voice, postal, or overnight delivery, making [13] the Net a very inexpensive communications medium. It is also a very fast method of communication, with messages arriving anywhere in the world in a matter of seconds or a minute or two at most. We will now briefly describe the most important Internet capabilities.

route[ru:t]v.（按一定的线路）传递,发送

The Internet is based on *client/server technology*. *Individuals* using the Net control what they do through client applications, using *graphical user interfaces* or *character-character products* that control all functions. All the data, including e-mail messages, databases, and Web sites, are stored on servers. Servers dedicated to the Internet or even to specific Internet functions are the heart of the information on the Net.

dedicate['dedikeit]v. 把…奉献给

The most important Internet capabilities for business include *e-mail*, *Usenet newsgroups*, *LISTSERVs*, *chatting*, *Telnet, FTP, gophers*, and *the World Wide Web*. They can be used to retrieve and offer information. Table 11.1 lists these capabilities and describes the functions they support.

retrieve[ri'tri:v]v.检索（存储的信息）
offer['ɔfə]v.提供;供应

Table 11.1   Major Internet Capabilities

Capability	Functions Supported
E-mail	Person-person messaging; document sharing
Usenet newsgroups	Discussion groups on electronic bulletin boards
LISTSERVs	Discussion groups using e-mail mailing list servers
Chatting	Interactive conversations
Telnet	Log on to one computer system and do work on another
FTP	Transfer files from computer to computer
Gophers	Locate information using a hierarchy of menus
World Wide Web	Retrieve, format, and display information(include text, audio, graphics, and video) using hypertext links

[11] Tip:Here the relative pronoun "which" refers to "regional Internet companies", and the subject of the relative clause is "member networks".
[12] Forward(verb.) sth to sb| forward sb sth:to send or pass goods or information to sb;发送（商品或信息）。
[13] Make—[VN -N]。

productivity[ˌprɔdʌk'tiviti]*n.*生产率;生产效率

facilitate[fə'siliteit]*v.*促进;使便利

facility[fə'siliti]*n.*特色服务

participant[pɑ:'tisipənt]*n.*参与者;参加者

Electronic Mail(E-Mail). The Net has become the most important e-mail system in the world because it connects so many people worldwide[14], creating a productivity gain that observers have compared to Gutenberg's development of movable type in the fifteenth century. Organizations use it to facilitate communication between employees and offices, and to communicate with customers and suppliers.

Researchers use this facility to share ideas, information, even documents. E-mail over the Net also has made possible many collaborative research and writing projects, even though the participants are thousands of miles apart[15]. With proper software, the user will find[16] it easy to attach documents and multimedia files when sending a message to someone or to broadcast a message to a predefined group. Figure 11.1 illustrates the components of an Internet *e-mail address.*

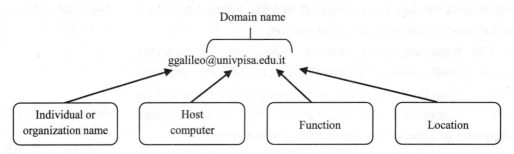

Figure 11.1   Analysis of an Internet address. In English, the e-mail address of physicist and astronomer Galileo Galilei would be translated as "G. Galileo @ University of Pisa, educational institution, Italy". The domain name to the right of the @ symbol contains a country indicator, a function indicator, and the location of the host computer.

portion['pɔ:ʃən]*n.*（一）部分

The portion of the address to the left of the @ *symbol* in Net e-mail address is the *name* or *identifier* of the specific individual or organization. To the right of the @ symbol is the domain name. The *domain name* is the unique name of a collection of computers connected to the Internet. The domain contains *subdomains* separated by a period. The domain that is farthest to the right is the *top level domain*, and each domain to the left helps further define the domain by network, department,

---

[14] Worldwide(adj.):affecting all parts of the world;影响全世界的；世界各地的。
[15] Apart(adv.):separated by a distance, of space or time;（指空间或时间）相隔；相距。
[16] Find—[VN -ADJ]。

and even specific computer. The top level domain name may be either a country indicator or a function indicator, such as *com* for a commercial organization or *gov* for a government institution. All e-mail address end with a *country indicator* except those in the United States, which ordinarily does not use one. In Figure 11.1, it, the top level domain, is a country indicator, indicating that the address is in Italy. *Edu* indicates that the address is an educational institution; univpisa(in this case, University of Oisa) indicates the specific location of the host computer.

Usenet Newsgroups(Forums). Usenet newsgroups are worldwide discussion groups in which people share information and ideas on a defined topic such as radiology or rock bands[17]. Discussion takes place in large *electronic bulletin boards* where anyone can post messages for others to read. Almost 20,000 groups exist discussing almost all conceivable topics. Each Usenet site is financed and administered independently.

LISTSERV. A second type of public forum, LISTSERV, allows discussions to be conducted through predefined groups but uses *e-mail mailing list servers* instead of[18] bulletin boards for communications. If you find a LISTSERV topic you are interested in, you may describe. From then on[19], through your e-mail, you will receive all messages sent by others concerning that topic[20]. You can, in turn, send a message to your LISTSERV and it will automatically be broadcast to the other subscribers. Tens of thousands of LISTSERV groups exist.

Chatting. Chatting allows two or more people who are simultaneously connected to the Internet to hold live, interactive conversations. *Internet Relay Chat(IRC)* is a general chat program for the Internet. *Chat groups* are divided into *channels*, and each is assigned its own topic of conversation. Most chat tools today are for written conversations in which participants type their remarks using their keyboard and read responses on their computer screen. However, systems featuring voice chat

indicator['indi,keitə] *n.*标识

radiology[,reidi:'ɔləd ʒi:]*n.*放射学;放射医疗

conceivable[kən'si:v əbl]*adj.*能想到的;可想象的

subscriber[səb'skiaibə] *n.*消费者;用户

simultaneously[saim əl'teiniəsli]*adv.*可同时进行地;同步地

---

[17] Rock band:摇滚乐队。

[18] Instead of(prep.):in the place of sb/sth;代替；作为…的替换。

[19] From ... on:starting at the time mentioned and continuously after that;从…时起。

[20] Tip:...all messages sent by others concerning that topic.

capabilities, such as those offered by Tribal Voice and by Mpath at its HearMe.com Web site are becoming available. Chatting can be an effective business tool if people who can benefit from interactive conversations set an appointed time to "meet" and "talk" on a particular topic. Many on-line retailers are enhancing their Web sites with chat servers to attract visitors, to encourage repeat[21] purchases, and to improve customer service.

purchase['pə:tʃəs]*n.*
购买;消费

Telnet. Telnet allows someone to be on one computer system while doing work on another. Telnet is the protocol that established an error-free,rapid link between the two computers, allowing you, for example, to log on[22] to your business computer from a remote computer when you are on the road or working from your home. You can also log in and use third-party computers that have been made accessible to the public, such as using the catalog of the U.S. Library of Congress. Telnet will use the computer address you supply to locate the computer you wan to reach and connect you to it[23].

Information retrieve is a second basic Internet function. Many hundreds of library catalogs are on-line through the Internet, including those of such giants as the library of Congress, the University of California, and Harvard University. In addition, users are able to search many thousands of databases that have been opened to the public by corporations, governments, and nonprofit organizations. Individuals can gather information on almost any conceivable topic stored in the databases and libraries. Many use the Internet to locate and download some of the free, quality computer software that has been available by developers on computers on computers all over the world.

corporation[ˌkɔ:pə're
iʃən]*n.*（大）公司
nonprofit[nɔn'prɔfit]
*adj.*非营利的

The Internet is a voluntary, decentralized effort with no central listing of participants or sites, much less a listing of the data located at those sites, so a major problem is finding what you need from among the storehouses of data found in databases and libraries. Here we introduce two major methods of accessing computers and locating files.

[21] Repeat(noun.):an event that is very similar to sth that happened before;重复的事件。
[22] Log in/on:to perform the actions that allow you to begin using a computer system;登录；注册；进入（计算机系统）。
[23] Tip:In this case, the relative pronoun is omitted!

FTP. *File transfer protocol(FTP)* is used to access a remote computer and retrieve files from it. FTP is a quick and easy method if you know the remote computer site where the file is stored. After you have logged on to the remote computer, you can move around directories that have been made[24] accessible for FTP to search for the file(s) you want to retrieve. Once located, FTP makes transfer of[25] the file to your own computer very easy.

Gopher. Most files and digital information that are accessible through FTP also are available through gophers. A gopher is a computer client tool that enables the user to locate information stored on Internet gopher servers through a serious of easy-to-use, hierarchical menus. The Internet has thousands of gopher server sites throughout the world. Each gopher site contains its own system of menus listing subject-matter topics, local files, and other relevant gopher sites. One gopher site might have as many as several thousand listings within its menus. When you use gopher software to search a specific topic and select a related item from a menu, the server will automatically transfer you to the appropriate file on that server or to the selected server wherever it is located. Once on that server, the process continue; you are presented with more menus of files and other gopher site servers that might interest you. You can move from site to site, narrowing your search as you go, locating information anywhere in the world. With descriptive menu listings linked to other gopher sites, you do not need to know in advance where relevant files are stored or the exact FTP address of a specific computer.

TERMINOLOGY

@ symbol   graphical user interface
channel   identifier

---

[24] Make—[VN -ADJ]。

[25] Of:used after nouns formed from verbs. The noun after "of" can be either the object or the subject of the action;用于由动词转化的名词之后，of 之后的名词可以是受动者，也可以是施动者。

character-character product	individual
chat group	information retrieve
chatting	Internet Relay Chat(IRC)
client/server technology	Internet Service Provider(ISP)
com	LISTSERV
country indicator	modem
domain name	name
edu	online service
electronic bulletin board	subdomain
e-mail	TCP/IP(Transmission Control Protocol/Internet Protocol)
e-mail address	Telnet
e-mail mailing list server	the Internet
file transfer protocol(FTP)	the Net
FTP	the World Wide Web
gopher	top level domain
gov	Usenet newsgroup

EXERCISES

11.1 Translate each of the following key terms:

    a)Internet Service Provider
    b)e-mail address
    c)client/server technology
    d)domain name
    e)the World Wide Web
    f)modem
    g)chat group
    h)the Internet

11.2 Fill in the blanks in each of the following statements:

    a)The largest network in the world is called the _____.
    b)_____ are the heart of the information on the Net.
    c)A(n) _____ provides access to the Internet.
    d)The _____ name is the unique name of a collection of computers connected to the Internet.
    e)A domain consists of _____ separated by a period.
    f)Chat groups are divided into _____ each of which is assigned its own topic

conversation.

g)Usenet newsgroup discussion takes place in _____.

h)The standard protocol for the Internet is _____.

i)_____ is an Internet service for uploading and downloading files.

j)Modulation and demodulation are the process of a(n) _____.

k)_____ is an Internet service that provides terminal access to host computer.

l)The function indicator _____ stands for a commercial organization.

m)A top level domain name is either a _____ or a function indicator.

n)The portion of a Net e-mail address to the left of the @ is the name or _____ of a specific individual or organization.

11.3 State whether each of the following is *true* or *false*. If *false*, explain why.

a)Anyone with a computer, a modem, and paying a monthly usage fee can access directly to the Internet.

b)The Internet is based on client/server technology.

c)All e-mail address end with a country indicator except those in the United States.

d)One popular chat service is called Internet Relay Chat.

e)Each Usenet site is financed and administered dependently.

f)A gopher is a computer client tool that enables the user to locate information stored on Internet gopher servers through a serious of menus.

g)Chatting allows two or more people who are simultaneously connected to the Internet to hold live, interactive conversations.

h)LISTSERV uses bulletin boards for communication.

i)The domain that is nearest to the left is the top level domain.

11.4 Match each numbered item with the most closely related lettered item:

a)e-mail	1.discussion groups using e-mail mailing list servers.
b)Usenet newsgroups	2.log on to one computer system and do work on another.
c)LISTSERVs	3.transfer files from computer to computer.
d)chatting	4.locate information using a hierarchy of menus.
e)Telnet	5.discussion groups on electronic bulletin boards.
f)FTP	6.retrieve, format, and display information(include text, audio, graphics, and video) using hypertext links.
g)gophers	7.person-person messaging; document sharing.
h)World Wide Web	8.interactive conversations.

11.5 Expand each of the following acronyms:

    a)ISP.
    b)TCP/IP.
    c)FTP.
    d)WWW.
    e)IRC.
    f)HTML.
    g)URL.
    h)HTTP.

11.6 Explain the components of the email address:
kermit@animals.com

Reading Material（阅读材料）

The World Wide Web

The World Wide Web(also referred to as WWW, W3, or the Web) is at the heart of the explosion in the business use of the Net. *The Web* is a system with universally accepted standards for storing, retrieving, formatting, and displaying information using a *client/server architecture*. It was developed to allow collaborators in remote sites to share their ideas on all aspects of a common project. If the Web was used for two independent projects and later relationships were found between the projects, information could flow smoothly between the projects without making major changes.

The Web combines text, hypermedia, graphics, and sound, by which *multimedia information* is disseminated over the Internet. It can handle all types of digital communication while making it easy to link resources that are half-a-world apart. The Web uses *graphical user interfaces* for easy viewing. It is based on a standard hypertext language called *Hypertext Markup Language(HTML)*, which formats documents and incorporates dynamic links to other documents and pictures stored in the same or remote computers. Using these links, the user need only point at a *highlighted* key word or graphic, click on it, and immediately be transported to another document, probably on another computer somewhere else in the world. Users are free to jump from place to place following their own logic and interest.

In this manner, a reader of hypertext documents can explore related documents or follow a train of thought from document to document. As portions of various documents are linked to other documents, an intertwined web of related information is formed.

When implemented on a computer network, the documents within such a web can reside on different machines, forming a network-wide web. The Web had its origins in the work of Tim Berners-Lee who realized the potential of combining the linked-document concept with internet technology and produced the first software for implementing the Web in December of 1990.

Software packages that allow users to access hypertext on the Internet fall into one of two categories:packages that play the role of clients, and packages that play the role of servers. A *client package* resides on the user's computer and is charged with the tasks of obtaining materials requested by the user and presenting these materials to the user in an organized manner. It is the client that provides the user interface that allows a user to browse within the Web. Hence the client is often referred to as a *browser*, or sometimes as a *Web browser*. The *server package*(often called a *Web server*) resides on a computer containing hypertext documents to be accessed. Its task is to provide access to the documents on its machine as clients. In summary, a user gains access to hypertext documents by means of a browser residing on the user's computer. This browser, playing the role of a client, obtains the documents by soliciting the services of the Web servers scattered throughout the Internet.

Web browser software is programmed according to HTML standards. The standard is universally accepted, so anyone using a browser can access any of the millions of Web sites. Browsers use hypertext's *point-and-click* ability to navigate or surf—move from site to site on the Web—to another desired site. The browser also includes an *arrow* or *back button* to enable the user to retrace his or her steps, navigating back, site by site.

Those who offer information through the Web must establish a *home page*—a text and graphical screen display that usually welcomes the user and explains the organization that has established the page. For most organizations, the home page will lead the user to other pages, with all the pages of a company being known as a *Web site*. For a corporation to establish a presence on the Web, therefore, it must set up a Web site of one or more pages. Most Web pages offer a way to contact the organization or individual. The person in charge of an organization's Web site is called a *Webmaster*.

To access a Web site, the user must specify a *uniform resource locator(URL)*, which point to the address of a specific resource on the Web. For instance, the URL for Prentice Hall, a world-famous publisher, is

http://www.prenhall.com

Http stands for *hypertext transport protocol*, which is the communications standard used to transfer pages on the Web. HTTP defines how messages are formatted and transmitted and what actions Web servers and browsers should take in response to various commands. www.prenhall.com is the *domain name* identifying the Web server storing the Web pages.

Locating information on the Web is a critical function given the tens of millions of Web sites in existence and growth estimated at 300,000 pages per week. No comprehensive catalog of Web sites exists. The principle methods of locating information in the Web are *Web site directories*, search engines, and broadcast or "push" technology.

Several companies have created directories of Web sites and their addresses, providing search tools for finding information. Yahoo! is an example. People or organizations submit sites of interest, which then are classified. To search the directory, you enter one or more keywords and will see displayed a list of categories and sites with those keywords in the title(see Figure 11.2).

Figure 11.2   Yahoo! provides a directory of Web sites classified into categories and is a major Internet portal. Users can search for sites of interest by entering keywords or exploring the categories.

Other search tools do not require Web sites to be preclassified and will search Web pages on their own automatically. Such tools, called *search engines*, can find Web sites that may be little known. They contain software that looks for Web pages containing one or more of the search terms; then it displays matches ranked by a method that usually involves the location and frequency of the search terms. These search engines do not display information about every site on the Web, but they create indexed of the Web pages they visit. The search engine software then locates Web pages of interest by searching through these indexes. Alta Vista, Lycos, and Infoseek are examples of these search engines. Some are more comprehensive or current than others, depending on how their components are tuned. Some also classify Web sites by subject categories. Specialized search tools are also available to help users locate specific types of information easily. For example, Google is tuned to find the home pages of companies and organizations.

Some Web pages for search engines such as Yahoo! and Lycos have become so popular and easy to use that they also serve as portals for the Internet. A *portal* is a Web site or other service providing an initial point of entry to the Web. Portals typically offer a broad array of resources or services such as e-mail, on-line shopping, discussion forums, and tools for locating information.

Instead of spending hours surfing the Web, users can have the information they are interested in delivered automatically to their desktops through *"push" technology*. A computer broadcasts information of interest directly to the user, rather than having the user "pull" content from Web sites.

"Push" comes from *server push*, a term used to describe the streaming of Web page contents from a Web server to a Web browser. Special client software allows the user to specify the categories of information he or she wants to receive, such as news, sports, financial data, and so forth, and how often this information should be updated. The software runs in the background of the user's computer while the computer performs other tasks. When they find the kind of information requested, push programs server it to the push client, notifying him or her by sending e-mail, playing a sound, displaying an icon on the desktop, sending full articles or Web pages, or displaying headlines on a screen saver. The stream of information distributed through push technology are also known as *channels* and can include private intranet channels and extranet channels, as well as channels from the public Internet. Microsoft's Internet Explorer and Netscape Communicator include push tools that automatically download Web pages, inform the user of updated content, and create channels of user-specified sites. The use of push technology to transmit information to a select group of individuals in one example of *multicasting*.

The audience for push technology is not limited to individual users. Companies are using push technology to set up their own channels to broadcast important internal information via corporate intranets or extranets. For example, Fruit of the Loom is using Pointcast push technology to alert managers to updated inventory information stored on its IBM AS/400 internet Web server. The company has long production schedules and a compressed selling season. When production of an item is behind schedule, warnings can be pushed to sales planners so they can contact customers or adjust promotions.

Note

Lesson 12　Network Security

When a computer is connected to a network, it becomes subject[1] to unauthorized access and vandalism. There are numerous ways that a computer system and its contents can be attacked via[2] network connections. Many of these incorporate the use of malicious software(collectively called *malware*[3]). Such[4] software might be transferred to, and executed on, the computer itself, or it might attack the computer from a distance. Examples of software that is transferred to, and executed on, the computer under attack include viruses, worms, Trojan horses, and spyware, whose names reflect the primary characteristic of the software.

A *virus* is software that infects a computer by inserting itself into programs that already reside in the machine. Then, when the "host" program is executed, the virus is also executed. When executed[5], many virtues do little more than try to transfer themselves to other programs within the computer. Some viruses, however, perform devastating actions such as degrading portions of the operating system, erasing large blocks of mass storage, or otherwise corrupting data and other programs.

A *worm* is an autonomous program that transfers itself through a network, taking up residence in computers and forwarding copies of itself to other computers. As in the case of a virus, a worm can be designed merely to replicate itself or to perform more extreme vandalism. A characteristic consequence of a worm is an explosion of[6] the worm's replicated copies that degrades the performance of legitimate applications and can ultimately overload an entire network or Internet.

vandalism['vændlˌɪzəm]*n.*恣意破坏;故意破坏
attack[ə'tæk]*v.*攻击;袭击
incorporate[in'kɔ:pəreit]*v.*包含;包括
malicious[mə'liʃəs]*adj.*恶意的;恶毒的

infect[in'fekt]*v.*传染,使感染（计算机病毒）
reside[ri'zaid]*v.*居住在;定居于

devastating['devəsteitiŋ]*adj.*毁灭性的
corrupt[kə'rʌpt]*v.*引起（计算机文件等的）错误;破坏
autonomous[ɔ:'tɔnəməs]*adj.*自治的
forward['fɔ:wəd]*v.*发送;转寄
replicated['repliˌkeit]*adj.*再生的;自我复制的
legitimate[li'dʒitimit]*adj.*正当的;合理的

---

[1] Subject(adj.) to sth:likely to be affected by sth, especially sth bad;可能受…影响的；易遭受…的。
[2] Via(prep.):by means of a particular person, system, etc.;通过，凭借（某人、系统等）。
[3] Malware:malicious software 的缩写。
[4] Such(det.):of the type already mentioned;（指上文）这样的，那样的。
[5] 同:When they are executed, ...
[6] Of:used after nouns formed from verbs. The noun after "of" can be either the object or the subject of the action;用于由动词转化的名词之后，of 之后的名词可以是受动者，也可以是施动者。

disguise[disˈgaiz]*v.* 伪
装;假扮
victim[ˈviktim]*n.*受害者;
受骗者

dormant[ˈdɔːmənt]
*adj.*潜伏的;蛰伏的
trigger[ˈtrigə]*v.*触发;
引起
enticing[inˈtaisiŋ]*adj.*
有诱惑力的;诱人的
misdeed[misˈdiːd]*n.*
恶行;不义之举

instigator[ˈinstigeitə]
n.（幕后）唆使者

blatantly[ˈbleitəntli]
adv.（坏的行为）明目张
胆地;公然地

explicitly[iksˈplisitli]
*adv.*不隐晦地;直截了
当地

con[kɔn]*n.*骗局;诡计
perpetrator[ˌpəːpiˈtre
itə]*n.*作恶者;犯罪者

A *Trojan horse* is a program that enters a computer system disguised as a desirable program, such as a game or a useful utility package[7], that is willingly imported by the victim. Once in the computer, however, the Trojan horse performs additional activities that might have harmful effects. Sometimes these additional activities start immediately. In other instances, the Trojan horse might lie[8] dormant until triggered by a specific event such as the occurrence of a pre-selected date. Trojan horses often arrive in the form of attachments to[9] enticing email messages. When the attachment is opened(that is, when the recipient asks to view the attachment), the misdeeds of the Trojan horse are activated. Thus, email attachment from unknown sources should never be opened.

Another form of malicious software is *spyware*(sometimes called *sniffing* software), which is software that collects information about activities at the computer on which[10] it resides and reports that information back to the instigator of the attack. Some companies use spyware as a means of building customer profiles, and in this context, it has questionable ethical merit[11]. In other cases, spyware is used for blatantly malicious purposes such as recording the symbol sequences typed at the computer's keyboard in search of passwords or credit card numbers.

As opposed to obtaining information secretly by sniffing via spyware, *phishing*[12] is a technique of obtaining information explicitly by simply asking for it. The term phishing is a play on the word fishing since the process involved is to cast numerous "lines" in hopes that someone will "take the bait." Phishing is often carried out via email, and in this form, it is little more than an old telephone con. The perpetrator sends email messages posing as a financial institution, a government bureau[13], or perhaps a law enforcement agency. The email asks the potential

---

[7] Utility packages:软件插件；实用程序包。

[8] Lie—[V -ADJ]。

[9] Attachment to sth:a document that you send to sb using email;（用电子邮件发送的）附件。

[10] Tip:Here the relative pronoun "which" refers to "the computer", and the subject of the relative clause is "it" referring to "spyware".

[11] 翻译：在这种情况下，遭到质疑的是它是否违背了道德标准。

[12] Tip:phishing 与 fish 同音，是 fish 的双关语。

[13] Bureau(noun.):(in the US) a government department or part of a government department;（美国政府部门）局；处；科。

victim for information that is supposedly needed for legitimate purposes. However, the information obtained is used by the perpetrator for hostile purposes.

In contrast to[14] suffering from such internal infections as viruses and spyware, a computer in a network can also be attacked by software being executed on other computers in the system. An example is a *denial of service attack*, which is the process of overloading a computer with requests. Denial of service attacks have been launched against large commercial Web servers on the Internet to disrupt the company's business and in some cases have brought the company's commercial activity to a halt.

A denial of service attack requires the generation of a large number of requests over a brief period of time. To accomplish this, an attacker usually plants software in numerous unsuspecting computers that will generate requests when a signal is given. Then, when the signal is given, all of these computers swamp the target with messages. Inherent[15], then, in denial of service attacks is the availability of unsuspecting computers to use as accomplices. This is why all PC users are discouraged from leaving their computers connected to the Internet, at least one intruder will attempt to exploit its existence within 20 minutes. In turn, an unprotected PC represents a significant threat to the integrity of the Internet.

Another problem associated with an abundance of[16] unwanted messages is the proliferation of unwanted junk emails, called *spam*. However, unlike a denial of service attack, the volume[17] of spam is rarely sufficient to overwhelm the computer system. Instead[18], the effect of spam is to overwhelm the person receiving the spam. This problem is compounded by the fact that, as we have already seen, spam is a widely adopted medium for phishing and instigating Trojan horses that

hostile['hɒstail]*adj.* 恶意的;敌对的

launch['lɔ:ntʃ]*v.*发动; 发起（尤指有组织的 活动）
disrupt[dis'rʌpt]*v.* 扰乱;中断
halt[hɔ:lt]*n.*暂停;停止

swamp[swɒmp]*v.* 使不堪承受;疲于应对

accomplice[ə'kɒmplis] *n.*从犯;帮凶

intruder[in'tru:də]*n.* 侵入者;闯入者

proliferation[prəu‚lifə 'reiʃən]*n.*（迅速）繁殖; 猛增
junk[dʒʌŋk]*n.*垃圾
overwhelm[‚əuvə'hw elm]*v.*压垮;使应接不暇
compound['kɒmpaund] *v.*使加重;使恶化

---

[14]　Contrast(noun.) to/with sb/sth:对比；对照。

[15]　Inherent(adj.) in sb/sth:that is a basic or permanent part of sb/sth and that cannot be removed;固有的；内在的。

[16]　Abundance[sing., U] of sth:a large quantity that is more than enough;大量。

[17]　Volume:the amount of sth;量；额。

[18]　Instead(adv.):in the place of sb/sth;代替；反而；却。

detrimental[ˌtetriˈme ntl]*adj.*有害的;不利的

adage[ˈædidʒ]*n.*谚语; 格言

terminate[ˈtə:mineit] *v.*（使）结束;终止

masquerade[ˌmæskə ˈreid]*v.*伪装;假扮

clandestine[klænˈdes tin]*adj.*暗中的;秘密的

might spread viruses and other detrimental software.

The old adage "an ounce of prevention is worth a pound of cure[19]" is certainly true in the context of controlling vandalism over network connections. A primary prevention technique is to *filter* traffic passing through a point in the network, usually with a program called a *firewall*. For instance, a firewall might be installed at a domain's gateway to filter messages passing in and out of the domain. Such firewalls might be designed to block outgoing messages with certain destination addresses or to block incoming messages from origins that are known to be sources of trouble. This latter function is a tool for terminating a denial of service attack since[20] it provides a means of blocking traffic from the attacking computers. Another common role of a firewall at a domain's gateway is to block all incoming messages that have origin addresses within the domain since such a message would indicate that an outsider is pretending to be a member of the domain. Masquerading as a party other than[21] one's self is known as *spoofing*.

Firewalls are also used to protect individual computers rather than[22] entire networks or domains. For example, if a computer is not being used a Web server, a name server, or an email server, then a firewall should be installed at that computer to block all incoming traffic addressed to such applications. Indeed, one way an intruder might gain entry to a computer is by establishing contact through a "hole" left by a nonexistent server. In particular, one method for retrieving information gathered by spyware is to establish a clandestine server on the infected computer by which malicious clients can retrieve the spyware's findings. A properly installed firewall could block the messages from these malicious clients.

Some variations of firewalls are designed for specific purposes—an example being *spam filter*, which are firewalls designed to block unwanted email. Many spam filters use rather sophisticated techniques to distinguish between desirable email

[19] 翻译：防病为主，治疗为辅。
[20] Since(conj.):because;as;因为；由于。
[21] Other than:different or in a different way from;not;不；非；不同于。
[22] Rather than:instead of sb/sth;而不是。

and spam. Some learn to make this distinction via a training process in which the user identifies items of spam until the filter acquires enough examples to make decisions on its own. These filters are examples of how a variety of subject ares(probability theory, artificial intelligence, etc.) can jointly contribute to developments in other fields.

Another preventative tool that has filtering connotations is the proxy server. A *proxy server* is a software unit that acts as a intermediary between a client and a server with the goal of shielding the client from adverse actions of the server. Without a proxy server, a client communicates directly with a server, meaning that the server has an opportunity to learn a certain amount about the client. Over time[23], as many clients within the same domain deal with a distant server, that server can collect a multitude of[24] information about the domain—information that can later be used to attack the domain. To counter this, a domain might contain a proxy server for a particular kind of server(FTP, HTTP, telnet, etc.). Each time a client within the domain tries to contact a server of that type, the client is actually placed in contact with the proxy server. Then the proxy server, playing the role of a client, contacts the actual server. Form then on the proxy server plays the role of an intermediary between the actual client and actual server by relaying messages back and forth[25]. The first advantage of this arrangement is that the actual server has no way of knowing that the proxy server is not the true client, and in fact, it is never aware of the actual client's existence. In turn, the actual server has no way of learning about the domain's internal features. The second advantage is that the proxy server is in position to filter all the messages sent from the server to the client. For example, an FTP proxy server could check all incoming file for the presence of known viruses and block all infected files.

Still another tool for preventing problems in a network environment is *auditing software* that is similar to the *utility packages.* Using network-level auditing software, a system

preventative/preven tive[pri'ventətiv]*adj.*
预防性的;防备的
connotation[ˌkɔnə'tei ʃən]*n.*含义;隐含意义
shield[ʃi:ld]*v.*保护;
庇护

---

[23] Over time:as time passes;随着时间的流逝。
[24] Multitude of sb/sth:an extremely large number of things or people;众多；大量。
[25] Back and forth:from one place to another and back again repeatedly;反复来回。

detect[di'tekt]*v.* 发现；
发觉

administrator can detect a sudden increase in message traffic at various locations within a domain, monitor the activities of the system's firewalls, and analyze the pattern of requests being made by the individual computers within the administrator's realm in order to detect irregularities. In effect, auditing software is an administrator's primary tool for identifying problems before they grow out of control.

irregularity[i,regjə'læriti:]*n.*不正当的行为；不正常的做法

invasion[in'veiʒən]*n.*侵入；侵犯

Another means of defense against invasions via network connection is software called *antivirus software*, which is used to detect and to remove the presence of known viruses and other infections. (Actually, antivirus software represents a broad class of software products, each designed to detect and remove a specific type of infection. For example, while many products specialize in virus control, other specialize in spyware protection.) It is important for users of these packages to understand that, just as in the case of biological systems, new computer infections are constantly coming on the scene that require updated vaccines. Thus, antivirus software must be routinely maintained by downloading *updates* from the software's vendor. Even this, however, does not guarantee the safety of a computer. After all[26], a new virus must first infect some computers before it is discovered and a vaccine is produced. Thus, a wise computer user never opens *email attachments* from unfamiliar sources, does not download software without first confirming its reliability, does not respond to *pop-up adds*[27], and does not leave a PC connected to the Internet when such connection is not necessary.

vaccine[væk'si:n]*n.*
疫苗
routinely[ru:'ti:nli]*adv.*
常规地；日常地

TERMINOLOGY

antivirus software

sniffing

auditing software

spam

---

[26] After all:despite what has been said or expected;毕竟；终归。
[27] Pop-up adds:弹出式广告。

denial of service attack	spam filter
email attachment	spoofing
filter	spyware
firewall	Trojan horse
malware	update
phishing	utility package
pop-up add	virus
proxy server	worm

EXERCISES

12.1 Translate each of the following key terms:

a)virus

b)antivirus software

c)firewall

d)proxy server

e)spam

f)spyware

g)Trojan horse

h)worm

12.2 Fill in the blanks in each of the following statements:

a)Another name for malicious software is _____.

b)A(n) _____ is an autonomous program that transfers itself through a network, taking up residence in computers and forwarding copies of itself to other computers.

c)Unwanted junk emails are also called _____.

d)A primary prevention technique of controlling vandalism over net is to _____ traffic passing through a point in the network.

e)_____ is software that collects information about activities at the computer on which it resides and reports the information back to the instigator of the attack.

f)A _____ might be installed at a domain's gateway to block outgoing messages with certain destination addresses or block incoming messages from origins that are known to be sources of trouble.

g)A software unit that acts as a intermediary between a client and a server with the goal of shielding the client from adverse actions of the server is a

_____.

h)_____ is used to detect and to move the presence of known viruses and other infections.

i)A(n) _____ proxy server could check all incoming file for the presence of known viruses and block all infected files.

j)_____ is a tool for preventing problems over net that is similar to the utility packages.

12.3 State whether each of the following is *true* or *false*. If *false*, explain why.

a)A characteristic consequence of a virus is an explosion of its replicated copies for malicious purposes.

b)Sometimes a Trojan horse might lie dormant until triggered by a specific event.

c)A denial of service attack is the process of overloading a computer with requests.

d)Worms often arrive in the form of attachments to enticing email messages.

e)The availability of unsuspecting computers to use as accomplices is inherent in denial of service attacks.

f)Some companies use sniffing software as a means of building customer profiles.

g)Phishing is a technique of obtaining information explicitly by asking for it via email for legitimate purposes.

12.4 Match each numbered item with the most closely related lettered item:

a)malware	1.firewall designed to block unwanted emails.
b)virus	2.program that enters a computer system disguised as a desirable program that is willing imported by the victim.
c)Trojan horse	3.also called sniffing software.
d)spyware	4.program that filters traffic passing in and out of a point in the network.
e)firewall	5.viruses, worms, Trojan horses, and spyware.
f)spam filter	6.software that infects a computer by inserting itself into programs that already reside in the machine.

12.5 What are two common ways that malware gains access to a computer system?

12.6 What distinction is there between the types of firewalls that can be placed at a domain's gateway as opposed to an individual host within the domain?

Reading Material（阅读材料）

Encryption

Increasingly people are using networks such as the Internet for on-line banking, shopping, and many other applications. The generic term used is electronic commerce or e-commerce and this often involves the transfer of *sensitive information* such as credit card details over the network. Hence to support this type of networked transaction, a number of security techniques have been developed which, when combined together, provide a high level of confidence that any information relating to the transaction that is received from the network:

- has not been altered in any way—*integrity*;
- has not been intercepted and read by anyone—*privacy/secrecy*;
- has come from an authorized sender—*authentication*;
- has proof that the stated sender initiated the transaction—*nonrepudiation*.

We shall describe a number of the techniques that are used to carry out these four functions. As we shall see, secrecy and integrity are achieved by means of *data encryption* while authentication and nonrepudiation require the exchange of a set of (encrypted) messages between the two communicating parties.

The traditional means of protecting information is to control its access through the use of passwords. However, passwords can be compromised and are of little value when data are transferred over networks and internets where messages are relayed by unknown entities. In these cases encryption can be used to so that even if the data fall into unscrupulous hands, the encoded information will remain confidential.

Data encryption(or *data encipherment*) involves the sending party—for example, the application protocol entity—in processing all data prior to transmission so that if it is accidentally or deliberately intercepted while it is being transferred it will be incomprehensible to the intercepting party. Of course, the data must be readily interpreted—*decrypted* or *deciphered*—by the intended recipient. Consequently, most encryption methods involve the use of an *encryption key*, which is hopefully known only by the two correspondents. The key features in both the encryption and the decryption correspondents. Prior to encryption, message data is normally referred to as *plaintext* and after encryption as *ciphertext*.

Today, many traditional Internet applications have been altered to incorporate data encryption techniques, producing what are called "secure versions" of the applications. Examples include *FTPS*, which is a secure version of *FTP(File Transfer Protocol)* and *SSH(Secure Shell)* as a secure replacement for telnet.

Still another example is the secure version of HTTP, known as *HTTPS*, which is used by most financial institutions to provide customers with secure Internet access to their account. The backbone of HTTPS is the protocol system known as *Secure Sockets Layer(SSL)*, which was originally developed by Netscape to provide secure communication links between Web clients and servers. Most browsers indicate the use of SSL by displaying a tiny padlock icon on the computer screen. (Some use the presence or absence of the icon to indicate whether SSL is being used, others display the padlock in either locked or unlocked position.)

One of the more fascinating techniques in the field of encryption is *public-key encryption*, which is an encryption system in which knowing how to encrypt messages does not allow one to decrypt message—a property that might seem counterintuitive. After all, intuition would suggest that a person that a person who knows how messages are encrypted should be able to reverse the process to decrypt messages. But, this is not true when public-key encryption techniques are used.

A public-key encryption system involves the use of two keys. One key, known as the *public key*, is used to encrypt messages; the other key, known as the *private key*, is required to decrypt messages. To use the system, the public key is first distributed to those who might need to send messages to a particular destination. The private key is held in confidence at this destination. Then, the originator of a message can encrypt the message using the public key and send the message to its destination with assurance that its contents are safe, even if it is handled by intermediaries who also know the public key. Indeed, the only party that can decrypt the message is the party at the message's destination who holds the private key. Thus if Bob creates a public-key encryption system and gives both Alice and Carol the public key, then both Alice and Carol can encrypt messages to Bob, but they cannot spy on the other's communication. Indeed, if Carol intercepts a message from Alice, she cannot decrypt it even though she knows how Alice encrypted it(Figure 12.1).

There are, of course, subtle problems lurking within public-key systems. One is to ensure that the public key being used is, in fact, the proper key for the destination party. For example, if you are communicating with your bank, you want to be sure that the public key you are using for encryption is the one for the bank and not an impostor. If an impostor presents itself as the bank(a process known as spoofing) and gives you its public key, the messages you encrypt and send to the "bank" would be meaningful to the impostor and not your bank. Thus, the task of associating public keys with correct parties is significant.

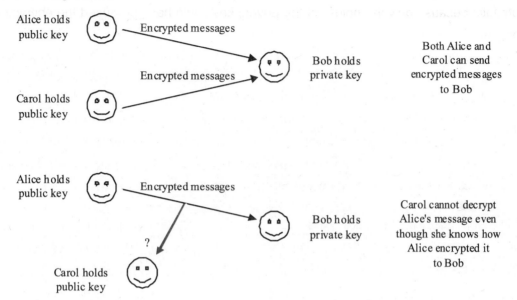

Figure 12.1   Public key encryption

One approach to resolving this problem is to establish trusted Internet sites, called *certificate authorities*, whose task is to maintain accurate lists of parties and their public keys. These authorities, acting as servers, then provides reliable public-key information to their clients in packages known as certificates. A *certificate* is a package containing a party's name and that party's public key. Many commercial certificate authorities are now available on the Internet, although it is also common for organizations to maintain their own certificate authorities in order to maintain tighter control over the security of the organization's communication.

Finally, we should comment on the role public-key encryption systems play in solving problems of *authentication*—making sure that the author of a message is, in fact, the party it claims to be. The critical point here is that, in some public-key encryption systems, the roles of the encryption and decryption keys can be revered. That is, text can be encrypted with the private key, and since only one party has access to that key, any text that is so encrypted must have originated from that party. In this manner, the holder of the private key can produce a bit pattern, called a *digital signature*, that only that party knows how to produce. By attaching that signature to a message, the sender can mark the message as being authentic to achieve *nonrepudiation*. A digital signature can be as simple as the encrypted version of the message itself. All the sender must do is encrypt the message being transmitted using his or her private key(the key usually used for decrypting). When the message is received, the receiver uses the sender's public key to decrypt the signature. The message that is revealed is guaranteed to be authentic because only the holder of the private key could have produced the encrypted version.

Note

Lesson 13 What Is an Information System?

interrelated[ˌintəriˈleitid]*adj.*相互关联的

retrieve[riˈtri:v]*v.*（计算机的）检索数据

visualize[ˈviʒuəlaiz]*v.*使形象化

An *information system* can be defined technically[1] as a set of interrelated components that collect(or retrieve), process, store, and distribute information to support *decision making* and *control* in an organization. In addition to[2] supporting decision making, *coordination*, and control, information systems may also help[3] managers and workers analyze problems, visualize complex subjects, and create new products.

Information systems contain information about significant people, places, and things within the organization or in the environment surrounding[4] it(Figure 13.1). By *information* we mean data that have been shaped into a form that is meaningful and useful to human beings. *Data*, in contrast[5], are streams of raw facts representing events occurring in organizations or the physical environment before they have been organized and arranged into a form that people can understand and use.

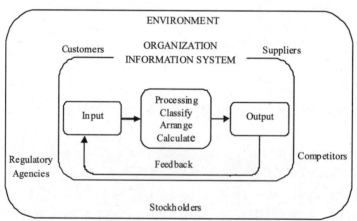

Figure 13.1   Functions of an information system. An information system contains information about an organization and its surrounding environment. Three basic activities—input, processing, and output—produce the information organizations need. Feedback is output returned to appropriate people or activities in the organization to evaluate and refine the input.

---

[1] Technically(adv.):in a way that is connected with the skills needed for a particular job, sport, art, etc.;在专业上；在技术上。

[2] In addition to sb/sth:used when you want to mention another person or thing after sth else;除…以外（还）。

[3] Help—[VN inf]。

[4] Tip:...in the environment surrounding it (Here "it" refers to "the organization".)

[5] Contrast(noun.):the fact of comparing two or more things in order to show the differences between them;对比；对照。

Three activities in an information system product the information that[6] organizations need for making decisions, controlling operations, analyzing problems, and creating new products or services. These activities are input, processing, and output. *Input* captures or collects *raw data* from within the organization or from its external environment. *Processing* converts this raw input into a more *meaningful form*. *Output* transfers the *processed information* to the people who will use it or to the activities for which[7] it will be used. Information systems also require *feedback*, which[8] is output that is returned to appropriate members of the organization to help them evaluate or correct the input stage.

capture['kæptʃə]*v.* 采集;收集

convert[kən'və:t]*v.* 使转变,转换

appropriate[ə'prəupriət]*adj.*相应的;恰当的
evaluate[i'væljueit]*v.*评价;评估
correct[kə'rekt]*v.*纠正;修正

In the information system used by InPart for obtaining parts, the raw input from designers and engineers consists of the type of component, manufacturer's name, part number, part size, or company's standard computerized design format specifications. The computer processes these data by locating components that match these criteria[9] from a massive computerized repository of parts data. Lists of components matching input criteria, three-dimensional representations of parts, online displays of pages from manufactures' catalogs, and customized computerized parts models become the output. The system thus provides meaningful information such as the right part model for a design, its dimensions and performance specifications, sources for the part models, part availability, and installation requirements.

massive['mæsiv]*adj.*大量的
repository[ri'pɔzi,tɔ:ri:]*n.*（正式）大仓库;大储物处

Our interest here is in formal, organizational *computer-based information systems(CBIS)* like those designed and used by InPart and its customers. *Formal systems* rest on[10] accepted and fixed definitions of data and procedures for collecting, storing, processing, disseminating, and using these data. The formal systems we describe in this text are structured;

fixed[fikst]*adj.*固定的;不变的
disseminate[di'semineit]*v.*传播,散布（知识、信息等）

---

[6] Tip:Here the relative pronoun "that" refers to "the information", and the subject of the relative clause is "organizations".
[7] 同:...the activities which it will be used for... (Here the relative pronoun "which" refers to "the activities", and "it" refers to "the processed information".)
[8] Tip:Note that we must include a relative pronoun in a non-defining relative clause!
[9] Criterion(pl. criteria):a standard or principle by which sth is judged, or with the help of which a decision is made; （评判或作决定的）标准, 准则, 原则。
[10] Rest on sth:to be based on sth;基于; 以…为基础。

conformity[kən'fɔ:mi ti]*n.*遵从,遵守

identify[ai'dentifai]*v.* 标识;显示;说明

that is, they operate in conformity with[11] predefined rules that are relatively fixed and not easily changed. For instance, InPart's system requires that orders for parts include the manufacture's name, and a unique number for identifying each part.

Informal information systems(such as *office gossip networks*) rely, by contrast, on unstated rules of behavior. These is no agreement on what is information, or on how it will be stored and processed. Such systems are essential for the life of an organization, but an analysis of their qualities is beyond the scope of this text.

manual['mænjuəl]*adj.* 手工的;手动的

Formal information systems can be either computer-based or manual. *Manual systems* use paper-and-pencil technology. These manual systems serve important needs, but they too are not the subject of this text. Computer-based information systems, in contrast, rely on computer hardware and software technology to process and disseminate information. From this point on[12], when we use the term information systems, we will be referring to computer-based information systems—formal organizational systems that rely on computer technology. The Window on Technology describes some of the typical technologies used in computer-based information systems today.

sharp[ʃɑ:p]*adj.*清晰的; 鲜明的

Although computer-based information systems use computer technology to process raw data into meaningful information, there is a sharp distinction between a computer and a computer program on the one hand[13], and an information system on the other. Electronic computers and related software programs are the technical foundation, the tools and materials, of modern information systems. Computers provide the equipment for storing and processing information. Computer programs, or software, are sets of *operating instructions* that direct and control computer processing. Knowing how computers and computer programs work is important in *designing solutions*

---

[11] In conformity with sth:following the rules of sth; conforming to sth;遵循（规则）；与…相符合（或一致）。

[12] From ... on:starting at the time mentioned and continuously after that;从…时起。

[13] On the one hand ... on the other (hand) ... :used to introduce different points of view, ideas, etc., especially when they are opposites; （引出不同的，尤指对立的观点、思想等）一方面…另一方面…。

to organizational problems, but computers are only part of an information system. Hosing provides an appropriate analogy. Houses are built with hammers, nails, and wood, but these do not make a house. The architecture, design, setting, landscaping, and all of the decisions that lead to the creation of these features are part of the house and are crucial for finding a solution to the problem of putting a roof over one's head. Computers and programs are the hammer, nails, and lumber of CBIS, but alone they cannot product the information a particular organization needs. To understand information systems, one must understand the problems they are designed to solve, their architectural and design elements, and the organizational process that lead to these solutions.

analogy[əˈnælədʒi]n.
类比;比拟
hammer[ˈhæmə]n.
锤子

TERMINOLOGY

computer-based information system(CBIS)
control
coordination
data
decision making
designing solution
feedback
formal system
information
information system

input
manual system
meaningful form
office gossip network
operating instruction
output
processed information
processing
raw data

EXERCISES

13.1 Translate each of the following key terms:

a)information system
b)data
c)information
d)input

e)output

f)feedback

g)computer-based information system

h)formal system

13.2 Fill in the blanks in each of the following statements:

a)An information system is a set of interrelated components to support _____, coordination, control, analysis, and visualization in an organization.

b)_____ is data that have been shaped into a form that is meaningful and useful to human beings.

c)An information system contains three basic activities:input, _____, output.

d)_____ are streams of raw facts representing events occurring in organizations or the physical environment.

e)_____ are sets of operating instructions that direct and control computer processing.

f)_____ is output returned to appropriate people or activities in a organization to evaluate and refine the input.

g)CBIS stands for _____.

13.3 Match each numbered item with the most closely related lettered item:

a)information system	1.the distribution of processed information to the people who will use it or to the activities for which it will be used.
b)information	2.system resting on accepted and fixed definitions of data and procedures, operating with predefined rules.
c)input	3.the collection of raw data from within the organization or from its external environment system for processing.
d)processing	4.interrelated components working together to collect, process, store, and disseminate information.
e)output	5.information systems that rely on computer hardware and software for processing and disseminating information.
f)CBISs	6.the conversion, manipulation, and analysis of raw input into a form that is more meaningful to humans.
g)formal system	7.processed data in a meaningful and useful form.

13.4 Distinguish between a computer, a computer program, and an information system. What is the difference between data and information?

13.5 What activities convert raw data to usable information in information systems? What is their relationship to feedback?

Reading Material（阅读材料）

The Challenge of Information Systems:Key Management Issues

Information systems today are creating many exciting opportunities for both businesses and individuals. They are also a source of new problems, issues, and challenges for managers. Here, you will learn about both the challenges and opportunities posed by information systems to enrich your learning experience.

Although information technology is advancing at a blinding pace, there is nothing easy or mechanical about building and using information systems. There are five key challenges confronting managers:

1. The Strategic Business Challenge:How can businesses use information technology to design organizations that are competitive and effective? Investment in information technology amounts to more than half of the annual capital expenditures of most large *service-sector firms*. Yet despite these heavy investments, many organizations are not obtaining significant business benefits. The power of computer hardware and software has grown much more rapidly than the ability of organizations to apply from information technology, many organizations actually need to be redesigned. They will have to make fundamental changes in *organizational behavior*, develop new *business models*, and eliminate the inefficiencies of outmoded *organizational structures*. If organizations merely automate what they are doing today, they are largely missing the potential of information technology. To fully benefit from information technology, including the opportunities provided by the Internet, organizations need to rethink and redesign the way they design, product, deliver, and maintain goods and services.

2. The Globalization Challenge:How can firms understand the business and system requirements of a global economic environment? The rapid growth in *international trade* and the emergence of a *global economy* call for information systems that can support both producing and selling goods in many different countries. In the past, each regional office of a *multinational corporation* focused on solving its own unique information problems. Given language, cultural, and political differences among countries, this focus frequently resulted in chaos and the failure of central management controls. To develop *integrated*, *multinational* information systems, businesses must develop global hardware, software, and communications standards and create *cross-cultural* accounting and reporting structures.

3. The Information Architecture Challenge:How can organizations develop an information architecture and information technology infrastructure that supports their business goals? Creating a new system now means much more than installing a new machine in the basement. Today, this process typically places thousands of terminals or personal computers on the desks of employees who have little experience with them, connecting the devices to powerful communications networks, rearranging social relations in the office and work locations, changing reporting patterns, and redefining business goals. Briefly, new systems today often require redesigning the organization and developing a new information architecture.

Information architecture is the particular form that information technology takes in an organization to achieve selected goals or functions. It is a design for the business application systems that serve each functional specialty and level of the organization and the specific way that they are used by each organization. Because managers and employees directly interact with these systems, it is critical for the success of the organization that its information architecture meet business requirements now and in the future.

The technology platform for this architecture is called the *information technology(IT) infrastructure* and consists of computer hardware, software, data and storage technology, networks, and human resources required to operate the equipment. These technologies constitute the *shared* IT resources of the firm and are available to all of its applications. Although this technology platform is typically operated by technical personnel, general management must decide how to allocate the resources it has assigned to hardware, software, data storage, and telecommunications networks to make sound information technology investments.

Here are typical questions regarding information architecture and IT infrastructure facing today's managers:Should the corporate sales data and function be distributed to each corporate remote site, or should they be centralized at headquarters? Should the organization purchase stand-alone personal computers or build a more powerful, centralized mainframe environment within an integrated telecommunications network? Should the organization build systems to connect the entire enterprise or *separate islands of applications*? There is no one right answer to these questions. Moreover, business needs are constantly changing, which requires the IT architecture to be reassessed continually.

Even under the best of circumstances, combining knowledge of systems and the organization is itself a demanding task. For many organizations, the task is even more formidable because they are crippled by fragmented and incompatible computer hardware, software, telecommunications networks, and information systems. Although Internet standards have solved some of these connectivity problems, integration of diverse computing platforms is rarely as *seamless* as promised. Many organizations are still struggling to integrate islands of information and technology into a coherent architecture.

4. The Information Systems Investment Challenge:How can organizations determine the business value of information systems? A major problem raised by the development of powerful, inexpensive computers involves not technology but management and organizations. It's one thing to use information technology to design, product, deliver, and maintain new products. It's another thing to make money doing it. How can organizations obtain a sizable payoff from their investment in information systems?

Engineering massive organizational and system changes in the hope of positioning a firm strategically is complicated and expensive. Is this an investment that pays off? How can you tell? Senior management can be expected to ask these questions:Are we receiving the kind of return on investment from our systems that we should be? Do our competitors get more? Understanding the costs and benefits of building a single system is difficult enough; it is daunting to consider whether the entire systems effort is "worth it". Imagine, then, how a senior executive must think when presented with a major transformation in information architecture—a bold venture in organizational change costing tens of millions of dollars and taking many years.

5. The Responsibility and Control Challenge:How can organizations design systems that people can control and understand? How can organizations ensure that their information systems are used in an ethically and socially responsible manner? Information systems are so essential to business, government, and daily life that organizations must take special steps to ensure that they are accurate, reliable, and secure. Automated or semiautomated systems that malfunction or are poorly operated can have extremely harmful consequences. A form invites disaster if it uses systems that don't work as intended, that don't deliver information in a form that people can interpret correctly and use, or that have control rooms where controls don't work or where instruments give false signals. The potential for massive fraud, error, abuse, and destruction is enormous.

Information systems must be designed so that they function as intended and so that humans can control the process. When building and using information systems, organizations should consider health, safety, job security, and social well being as carefully as they do their business goals. Managers will need to ask:Can we apply high quality assurance standards to our information systems, as well as to our products and services? Can we build information systems that respect people's rights of privacy while still pursuing our organization's goals? Should information systems monitor employees? What do we do when an information system designed to increase efficiency and productivity eliminates people's jobs?

Note

Lesson 14　Electronic Commerce and Electronic Business

conduct[kənˈdʌkt]v.
组织;安排

confront[kənˈfrʌnt]v.
使…无法回避;降临于…

underlying[ˌʌndəˈlaiiŋ]
adj.根本的;基础的

vast[vɑ:st]adj.巨大的;
广阔的

exchange[iksˈtʃeindʒ]
v.交换;互换

Creating new ways of conducting business electronically[1] both inside and outside[2] a firm is the most essential challenge confronting[3] *information systems*. Increasingly, *the Internet* is providing the underlying technology for the challenge. The Internet can link thousands of organizations into a single network, creating the foundation for a vast electronic marketplace. An electronic market is an information system that links together many buyers and sellers to exchange *information*, *products*, *services*, and *payments*. Through[4] computers and networks, these systems function like *electronic middlemen*, with lowered *costs* for typical *marketplace transactions* such as selecting suppliers, establishing prices, ordering goods, and paying bills. Buyers and sellers can complete purchase and sale transactions digitally, regardless of their location.

furiously[ˈfjuəriəsli]
adv.猛烈地;激烈地;疯
狂地

brochure[ˈbrəuʃuə]n.
资料（或广告）手册

manual[ˈmænjuəl]n.
使用手册;说明书

A vast array[5] of goods and services are being advertised, bought, and exchanged worldwide[6] using the Internet as a *global marketplace*. Companies are furiously creating eye-catching electronic brochures, advertisements, product manuals, and order forms on *the World Wide Web*. All kinds of products and services are available on *the Web*, including fresh flowers, books, real estate, musical recordings, electronics, and steaks.

mall[mɔ:l]n.购物广场;
大卖场

Many *retailers* maintain their own site on the Web, such as Virtual Vineyards, an on-line source of wine and food items. Others offer their products through *electronic shopping* malls, such as the Internet Shopping Network. *Customers* can locate products on this mall either by *manufacturer*, if they know what they want, or by *product type*, and then order them directly. Even

---

[1] Electronically:in an electronic way, or using a device that works in an electronic way, or using a computer;用电子方式；用电子装置；用计算机。

[2] Inside(prep.); outside(prep.).

[3] 同:...the changes that are confronting information systems.

[4] Through(prep.):by mens of; because of;以；凭借；因为；由于。

[5] Array[usually sing.]:a group or collection of things or people, often one that is large or impressive;大量。

[6] Worldwide(adv.).

electronic *financial trading* has arrived on the Web, offering electronic trading in stocks, bonds, mutual funds[7], and other financial instruments.

stock[stɔk]*n.*股票
bond[bɔnd]*n.*债券
mutual['mju:tʃuəl]*adj.*共有的;共同的

The Web is being increasingly used for *business-business transactions* as well. For example, airlines can use the Boeing Corporation's Web site to order parts electronically and check the status of their orders.

The global availability of the Internet for the exchange of[8] transactions between buyers and sellers is fueling the growth of electronic commerce. *Electronic commerce* is the process of buying and selling goods and services electronically with computerized business transactions using the Internet, networks, and other digital technologies. It also encompasses activities supporting those market transactions, such as advertising, marketing, customer support, delivery, and payment. By replacing manual and paper-based procedures with electronic alternatives, and by using *information flows* in new and dynamic ways, electronic commerce can accelerate ordering, delivery, and payment for goods and services while reducing companies' operating and inventory costs.

fuel[fjuəl]*v.*加强;刺激

encompass[en'kʌmpəs]*v.*包含;涉及（大量事物）

alternative[ɔ:l'tə:nətiv]*n.*替代物;替换品
accelerate[æk'seləreit]*v.*使加速;加快
inventory['invəntri]*n.*（商店的）库存;存货

The Internet is emerging as the primary technology platform for electronic commerce. Equally important, *Internet technology* is being increasing applied to facilitate the management of the rest of the business—publishing employee *personnel policies*, reviewing *account balances* and *production plans*, scheduling plant *requires* and *maintenance*, and revising *design documents*. Companies are taking advantage of the connectivity and ease of use of Internet technology to create internal corporate networks called *intranets* that are based on Internet technology. Use of these private intranets for organizational communication, collaboration, and coordination is soaring. In this text, we use the term *electronic business* to distinguish these uses of Internet and digital technology for the management and coordination of other business processes from electronic commerce.

facilitate[fə'siliteit]*v.*促进;促使;使便利

soar[sɔ:]*v.*猛增;急升

---

[7] Mutual fund=unit trust;单位信托投资公司（代客户进行不同组合的投资）。

[8] Of:used after nouns formed from verbs. The noun after "of" can be either the object or the subject of the action;用于由动词转化的名词之后，of之后的名词可以是受动者，也可以是施动者。

Table 14.1 Examples of Electronic Commerce and Electronic Business

Electronic Commerce
Amazon.com operates a virtual storefront on the Internet offering more than 3 million book titles for sale. Customers can input their orders via Amazon.com's Web site and have the books shipped to them.
Travelocity provides a Web site that can be used by consumers for travel and vacation planning. Visitors can find out information on airlines, hotels, vacation packages, and other travel and leisure topics, and they can make airline and hotel reservations on-line through the Web site.
Mobil Corporation created a private network based on Internet technology that allows its 300 lubricant distributors to submit purchase orders on-line.

Electronic Business
Roche Bioscience scientists worldwide use an intranet to share research results and discuss findings. The intranet also provides a company telephone directory and newsletter.
University of Texas Medical Branch at Galveston publishes nursing staff policies and procedures on an intranet. The intranet reduces paperwork and enhances the quality of nursing services by providing immediate notification of policy changes.
Dream Works SKG uses an intranet to check the daily status of projects, including animation objects, and to coordinate movie scenes.

By distributing information through electronic networks, electronic business extends the reach of existing management. *Managers* can use *e-mail*, *Web documents*, and *work-group software* to effectively communicate frequently with thousands of employees, and even to manage far-flung task forces and teams. These tasks would be impossible in face-to-face traditional organizations. Table 14.1 lists some examples of electronic commerce and electronic business.

far-flung['fɑː'flʌŋ]*adj.* 分布广的;广泛的

Figure 14.1 illustrates an enterprise making intensive use of Internet and digital technology for electronic commerce and electronic business. Information can flow seamlessly among different parts of the company and between the company and external entities—its *customers*, *suppliers*, and *business partners*. Organizations will move toward this vision as they increasingly use the Internet and networks to manager their internal processes and their relationship with customers, suppliers, and other external entities.

seamlessly['siːmlislɪ] *adv.*无缝地

ELECTRONIC BUSINESS   ELECTRONIC COMMERCE

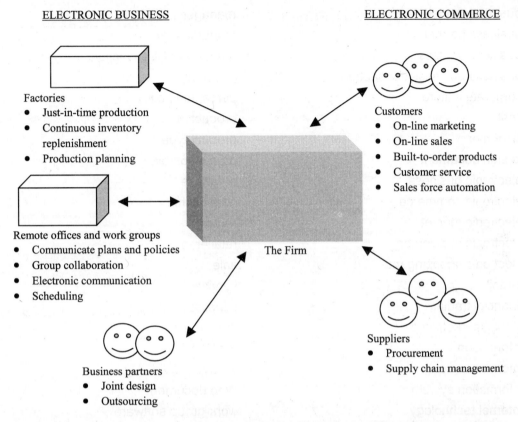

Factories
- Just-in-time production
- Continuous inventory replenishment
- Production planning

Remote offices and work groups
- Communicate plans and policies
- Group collaboration
- Electronic communication
- Scheduling

The Firm

Customers
- On-line marketing
- On-line sales
- Built-to-order products
- Customer service
- Sales force automation

Suppliers
- Procurement
- Supply chain management

Business partners
- Joint design
- Outsourcing

Figure 14.1   Electronic commerce and electronic business in the networked enterprise. Electronic commerce uses Internet and digital technology to conduct transactions with customers and suppliers, whereas electronic business uses these technologies for the management of the rest of the business.

Both electronic commerce and electronic business can fundamentally change the way[9] business is conducted. To use the Internet and other digital technologies successful for electronic commerce and electronic business, organizations may have to redefine their *business models*, reinvent *business processes*, change *corporate cultures*, and create much closer *relationships* with customers and suppliers.

redefine['riːdi'fain]*v.*
重新定义
reinvent[ˌriːin'vent]*v.*
以新形式出现

TERMINOLOGY

account balance   maintenance

---

[9] Tip:...the way business is conducted.

business model	manager
business partner	manufacturer
business process	marketplace transaction
business-business transaction	payment
corporate culture	personnel policy
cost	product
customer	product type
design document	production plan
electronic business	purchase
electronic commerce	relationship
electronic market	require
electronic middlemen	retailer
electronic shopping mall	sale
e-mail	service
financial trading	supplier
global marketplace	the Internet
information	the Web
information flow	the World Wide Web
information system	Web document
Internet technology	work-group software
intranet	

EXERCISES

14.1 Translate each of the following key terms:

    a)electronic market
    b)electronic commerce
    c)intranet
    d)electronic business
    e)financial trading
    f)retailer
    g)manufacturer
    h)global marketplace

14.2 Fill in the blanks in each of the following statements:

    a)A(n) _____ is a marketplace that is created by computer and communication technologies that link many buyers and sellers.

b)Electronic commerce is also known as _____.

c)Electronic commerce involves two parties:businesses and _____.

d)Three basic types of EC are B2C, _____, and B2B.

e)_____ are private networks within an organization that resemble the Internet.

f)B2B stands for _____.

14.3 Categorize each of the following examples as either electronic commerce or electronic business:

a)Amazon.com

b)Roche Bioscience

c)Dream Works SKG

d)Mobil Corporation

e)Travelocity

f)University of Texas Medical Branch at Galveston

14.4 Match each numbered item with the most closely related lettered item:

a)electronic commerce	1.offers electronic trading in stocks, bounds, mutual funds, and other financial instruments.
b)electronic market	2.Internet equivalent to traditional mall.
c)intranet	3.the use of the Internet and other digital technology for organizational communication and coordination and the management of the firm.
d)electronic business	4.similar to a traditional marketplace, but buyers and sellers interact only on the Web.
e)electronic financial trading	5.internal network based on Internet and World Wide technology and standards.
f)electronic shopping mall	6.the process of buying and selling goods and services electronically with computerized business transactions using the Internet, networks, and other digital technologies.

14.5 Expand each of the following acronyms:

a)EC.

b)EB.

c)EDI.

d)WAN.

e)VAN.

f)UPS.

g)DSP.

14.6 What is the relationship between the network revolution, electronic commerce, and electronic business?

Reading Material（阅读材料）

Customer-Centered Retailing & Business-to-Business Electronic Commerce

An array of information technologies are transforming the way products are produced, marketed, shipped, and sold. Companies have been using their own WANs, VANs, electronic data interchange(EDI), e-mail, shared databases, digital image processing, bar coding, and interactive software to replace telephone calls and paper-based procedures for product design, marketing, ordering, delivery, payment, and customer support. Trading partners can directly communicate with each other, bypassing middlemen and inefficient multilayered procedures. The Internet provides a public and universally available set of technologies for these purposes. It offers businesses an even easier way to link with other businesses and individuals at a very low cost. Web sites are available to consumers 24 hours a day. New marketing and sales channels can be created. Handling transactions electronically can reduce transaction costs and delivery time for some goods, especially those that are purely digital(such as software, text products, images, or videos). It is estimated that over $300 billion in goods and services will be exchanged over the Internet by 2002.

Just like any other type of commerce, electronic commerce involves two parties:*businesses* and *customers*. There are two major types of electronic commerce:*customer-centered retailing* and *business-to-business electronic commerce*.

Customer-Centered Retailing

The Internet provides companies with new channels of communication and interaction that can create closer yet more-effective relationships with customers in sales, marketing, and customer support.

a. Direct Sales over the Web

Manufacturers can sell their products and services directly to retail customers, bypassing intermediaries such as distributors or retail outlets. Eliminating middlemen in the distribution channel can significantly lower purchase transaction costs. Operators of *virtual storefronts* such as the Amazon.com on-line bookstore or

Virtual Vineyards do not have expenditures for rent, sales staff, and the other operations associated with a traditional retail store. Airlines can sell tickets directly to passengers through their own Web sites or through *travel sites* such as Travelocity without paying commissions to travel agents.

To pay for all the steps in a traditional distribution channel, a product may have to be priced as high as 135 percent of its original cost to manufacture. Figure 14.2 illustrates how much saving can result from eliminating each of these layers in the distribution process. By selling directly to consumers or reducing the number of intermediaries, companies can achieve higher profits while charging lower prices. The removal of organizations or business process layers responsible for intermediary steps in a value chain is called *disintermediation*.

The Internet is accelerating disintermediation in some industries and creating opportunities for new types of intermediaries in others. In certain industries, distributors with warehouses of goods, or middlemen such as real estate agents may be replaced by new intermediaries specializing in helping Internet users efficiently obtain product and price information, locate on-line sources of goods and services, or manage or maximized the value of the information captured about them in electronic commerce transactions. In businesses impacted by the Internet, middlemen will have to adjust their services to fit the new *business model* or create new services based on the model.

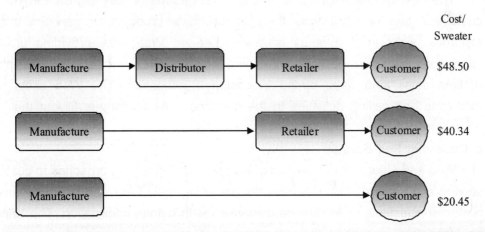

Figure 14.2   The benefits of disintermediation to the consumer. The typical distribution channel has several intermediary layers, each of which adds to the final cost of a product, such as a sweater. Removing layers lowers the final cost to the consumer.

b. Interactive Marketing

Marketers can use the interactive features of *Web pages* to hold consumers' attention or to capture information about their tastes and interests. Some of this information may be obtained by asking visitors to "*register*" on-line and provides information about themselves. Companies also can use special Web site auditing software capable of tracking the number of hits to their Web sites and the Web pages of greatest interest to visitors after they have entered the sites. (A *hit* is an entry into the log file of a Web server generated by each request to the server for a file.) They can analyze this information to develop more precise profiles of their customers. For instance, TravelWeb, a Web site offering electronic information on more than 16,000 hotels in 138 countries and on-line reservation capability, tracks the origin of a resort, valuable information in shaping market strategies and for developing hospitality-related products.

Companies can even use the Web and Internet capabilities such as electronic discussion groups, mailing lists, and e-mail to create ongoing dialogues with their customers. For example, visitors to the Web site for Reebok International Ltd. can obtain profiles of athletes and training tips from coaches in fitness categories of their preference. If they fill out profile forms that ask them to list their favorite sports, they also will receive customized workout tips, news updates about their sports, and other information on future visits. By becoming site "members", they can send e-mail "postcard" to their favorite athletes. Reebok further enhanced its Web site with e-mail, *discussion-group*, and *bulletin-board* capabilities to create a *community* of users.

The cost of customer surveys and focus groups is very high. Learning how customers feel or what they think about one's products or services through electronic visits to Web sites is much cheaper. Web sites providing products information also lower costs by shortening the sales cycle and reducing the amount of time sales staff must spend in customer education. The Web shifts more marketing and selling activities to the customers, as customers fill out their own *on-line order forms*.

c. Customer Self-Service

The Web and other network technologies are inspiring new approaches to customer service and support. Many companies are using their Web and e-mail to answer customer questions or to provide customers with helpful information. The Window on technology describes the experiences of companies using *chat rooms* and *Web-linked call centers* for this purpose.

The Web provides a medium through which customers can interact with the company, at their convenience, and find information on their own that previously required a human customer-support expert. Some companies are realizing substantial cost saving from *Web-based customer self-service applications*. American, Northwest, and other major airlines have created Web sites where customers can review flight departure and arrival times, seating charts, and airport logistics, check frequent-flyer miles, and purchase tickets on-line.

These Web sites allow companies to engage in ongoing dialogues with their customers that can provide information for other purposes. For example, Dell Computer has established a Dell *newsgroup* on the Net and other on-line services to receive and handle customer complaints and questions. They answer about 90 percent of the questions within 24 hours. Dell also does market research for free through these newsgroups rather than paying a professional for the same information.

Business-to-Business Electronic Commerce:New Efficiencies and Relationships

Many believe that the most promising area of electronic commerce is not retailing to individuals but the automation of purchase and sale transactions from *business to business*. For a number of years, companies have used proprietary *electronic data interchange(EDI)* systems for this purpose; now they are turning to the Web and *extranets*. Cisco systems, a leading manufacturer of networking equipment, conducts 40 percent of its sales electronically, with more than $1 billion in sales per year through its Web site. Order-taking, credit checking, production scheduling, technical support, and routine customer-support activities are handled on-line.

Marshall Industries' Virtual Distribution System

Marshall Industries, the world's fourth-largest distributor of industrial electronic components and production supplies, created a "virtual" distribution environment in which almost all of the processes it performed physically have been converted to a digital service on the Net.

Marshall's customers and suppliers can access its *intranet* to obtain customized information. For example, high-tech suppliers can see information about their own accounts, such as sales reports, inventory levels, and design data. They also can accept or reject price quotes or order training materials. A personal knowledge-assistant process called *Plugged-In* allows customers to specify the product categories they are interested in. They only received information specific to their interests.

When a customer places an order, the system verifies price and quantity and initiates a real-time credit authorization and approval. As soon as the order is approved, the system sends an automated request to the warehouse for scheduling. The system then sends the customer an order acknowledgement accompanied by relevant shipping and logistics information from UPS. Messages about order status are automatically "pushed" to the customer. The system thus integrates the entire process of placing and receiving an order.

Other features of Marshall's Web site provide additional service and value. Visitors can access a free "Electronic Design Center" to test and run their designs over the Internet. For example, an engineer might use Marshall's Web site to test Texas Instruments'(TI) *digital signal processors(DSPs)* for the design of a new piece of multimedia hardware. (DSP chips improve the performance of high-tech products such as computer hard disks, headphones, and power steering in cars.) At the site, the engineer would download sample code, modify the code to suit the product being built, test it on a "virtual chip" attached to the Web, and analyze its performance. If the engineer liked the results, Marshall could download his or her code, burn it into physical chips, and send back samples for designing prototypes. The entire process would take minutes.

Marshall's Web site provides after-sale training so that engineers do not have to attend special training classes or meetings in faraway locations. Marshall links to NetSeminar, a Web site where Marshall's customers can register for and receive educational programs developed for them by their suppliers using video, audio, and real-time chat capabilities.

For business-to-business electronic commerce, companies can use their own Web sites, like Cisco Systems and Marshall Industries, or they can conduct sales through Web sites set up as on-line marketplaces. Industrial malls such as Industrial Marketplace bring together a large number of suppliers in one place, providing search tools so that buyers can quickly locate what they need. They make money by collecting fees from their "tenant" vendors. Companies also can sell to other companies through Web sites that run cyberauctions for electronic parts and industrial and scientific equipment.

Corporate purchasing traditionally has been based on long-term relationships with one or two suppliers. The Internet makes information about alternative suppliers more accessible so that companies can find the best deal from a wide range of sources, including those overseas. For example, Mike Maiorano, the purchasing manager for XLNT Designs Inc., a manufacturer of networking technologies, consults the Web when he is asked to buy from an unfamiliar supplier or locate a new type of part. It is not surprising that identifying and researching potential trading partners is the most common procurement activity on the Internet. Suppliers themselves can use the Web to research competitors' prices on-line.

Organizations also can use the Web to solicit bids from suppliers by advertising requests for proposals on-line. Government organizations, especially military agencies and state governments, have been quick to adopt this model. Table 14.2 describes other examples of business-to-business electronic commerce.

Table 14.2　Examples of Business-to-Business Electronic Commerce

Business	Electronic Commerce Applications
U.S. General Services Administration	The procurement arm of the U.S. federal government created an ordering system called GSA Advantage, which allows federal agencies to buy everything through its Web site. The Web site lists 220,000 products and accounts for annual sales of $12 million. By using the Web, agencies can see all of their purchasing options and make choices based on price and delivery.
AMP Inc.	By placing its 400 catalogs on the Web, this electrical-components manufacturer hopes to reduce and eventually eliminate $8 million to $10 million per year in printing and shipping costs while offering catalogs that are always up-to-date. AMP created a new division called AMPeMerce Internet Solutions to help manufacturers and other companies develop Internet-based product catalogs and selling mechanisms.
General Electric Information Services	Operates a Trading Process Network(TPN) where GE and other subscribing companies can solicit and accept bids from selected suppliers over the Internet. TPN is a secure Web site developed for internal GE use that now is available to other companies for customized bidding and automated purchasing. GE earns revenue by charging subscribers for the service and by collecting a fee from the seller if a transaction is completed.

Note

Lesson 15　Elements of a 3D Game

The architecture of a modern *3D game* encompasses several discrete elements:the engine, scripts, GUI, models, textures, audio, and support infrastructure. In this section I'll give you some brief sketches of[1] each element to give you a sense of where we are going.

Game Engine

Game engines provide most of[2] the significant features of a *gaming environment*:3D *scene rendering*, *networking*, *graphics*, and *scripting*, to name[3] but a few. See Figure 15.1 for a block diagram that depicts the major feature areas.

User Input	Graphics	Audio
Event, Timing, & Synchronization	Scene Graph	Networking
Scripting	Objects & Resources	
File I/O		

Figure 15.1　Elements of a game engine

Game engines also allow for a sophisticated rendering of[4] *game environments*. Each game uses a different system to organize how the visual aspects of the game will be modeled. This becomes increasingly important as[5] games are becoming more focused on 3D environments, rich textures and forms, and an overall realistic feel to the game. *Textured polygon rendering* is one of the most common forms of rendering in *First-Person Shooter(FPS)* games, which tend[6] to be some of the more visually immersive[7] games on the market.

encompass[enˈkʌmpəs]v.包含;涉及（大量事物）

discrete[disˈkri:t]adj.分离的;互不相连的

sketch[stetʃ]n.概述;简述

significant[sigˈnifikənt]adj.重要的;有显著意义的

rendering[ˈrendəriŋ]n.渲染

depict[diˈpikt]v.描写;描述

sophisticated[səˈfistikeitid]adj.高级的;先进的

model[mɔdəl]v.建模

realistic[riəˈlistik]adj.真实的;逼真的

---

[1] Of:used after nouns formed from verbs. The noun after "of" can be either the object or the subject of the action;用于由动词转化的名词之后, of 之后的名词可以是受动者，也可以是施动者。

[2] Of:used to show sb/sth belongs to a group, often after *some, a few*,etc.;（常用在 some、a few 等词语之后，表示人或物的所属）属于…的。

[3] Name(verb.):to state exactly;说定；确定；准确描述。

[4] Of:used after nouns formed from verbs. The noun after 'of' can be either the object or the subject of the action;用于由动词转化的名词之后, of 之后的名词可以是受动者，也可以是施动者。

[5] As(conj.):used to state the reason for sth;因为；由于。

[6] Tend:to be likely to do sth or to happen in a particular way because this is what often or usually happens;往往会；常常是。--[V to inf]。

[7] Immersive(adj.):(technical) used to describe a computer system or image that seems to surrounding the user;（计算机系统或图像）沉浸式虚拟现实的。

obey[əu'bei]v.遵守;
遵循

narrative['nærətiv]n.
描述;叙述

constrain[kən'strein]
v.限制;约束

suspension[sə'spenʃ
ən]n.延迟;延缓

gravity['græviti]n.
重力;地心引力

scoring['skɔ:riŋ]n.
计算得分

combination[ˌkɔmbi'
neiʃən]n.联合体;综合体

By[8] creating consistent *graphic environments* and populating those environments with objects that obey specific physical laws and requirements, gaming engines allow games to progress significantly along the lines of producing more and more plausible narratives. *Characters* are constrained by rules that have realistic bases that increase the gamer's suspension of disbelief and draw him deeper into the game.

By including *physics formulas*, games are able to realistically account for [9] moving bodies, falling objects, and particle movement. This is how FPS games such as Tribes 2, Quake 3, Half-Life 2, or Unreal II are able to allow characters to run, jump, and fall in a virtual game world. Game engines encapsulate real-world characteristics such as time, motion, the effects of gravity, and other *natural physical laws*. They provide the developer with the ability to almost directly interact with the gaming world created, leading [10] to more immersive game environments.

Scripts

As you've just seen, the engine provides the code that[11] does all the hard work, graphics rendering, networking, and so on. We tie all these capabilities together with *scripts*. Sophisticated and fully featured games can be difficult to create without scripting capability.

Scripts are used to bring the different parts of the engine together, provide the game *play functions*, and enable the game world rules. Some of the things we will do with scripts include scoring, managing players, defining player and vehicle behaviors, and controlling GUI interfaces.

Graphical User Interface

The *Graphical User Interface(GUI)* is typically a combination of the graphics and the scripts that carries the visual appearance of[12] the game and accepts the user's *control inputs*. The player's *Heads Up Display(HUD)*, where health and score are displayed,

---

[8] Tip:We can use "by" with a present participle (-**ing**) clause with an adverbial meaning. This -**ing** clause indicates "the method or means used".

[9] Account for:to give an explanation of sth;解释；说明。

[10] Tip: ... the ability to almost directly interact with the gaming world created (leading to ...)

[11] Tip:In this case, the relative pronoun can't be omitted!

[12] Of:used after nouns formed from verbs. The noun after "of" can be either the object or the subject of the action;用于由动词转化的名词之后，of 之后的名词可以是受动者，也可以是施动者。

is part[13] of the GUI. So are the main start-up menus, the setting or option menus, the dialog boxes, and the various in-game message systems.

Figure 15.2 shows an example main screen using the Tubettiworld game. In the *upper-left corner*, the text that says "Client 1.62" is an example of a *GUI text control*. Stacked along the left side from the middle down are four *GUI button controls*. The popsicle-stick snapper *logo* in the lower right and the Tubettiworld logo across the top of the image are *GUI bitmap controls* that are overlayed on top of another GUI bitmap control(the background picture). Note that in the figure the top button control(Connect) is currently highlighted, with the *mouse cursor* over the top of it. This capability is provided by the *Torque Game Engine* as part of the definition of the button control.

overlay[ˌəuvəˈlei]v.
在…上;覆盖

Figure 15.2　An example of an main menu GUI

Models

3D *models*(Figure 15.3) are the essential soul of 3D games. With one or two exceptions, every visual item on a game screen that isn't part of the GUI is a model of some[14] kind. Our player's character is a model. The world[15] he tromps on is a

soul[səul]n.灵魂

tromp/tramp[trɔmp]
v.（尤指长时间地）重步
行走,踏,踩

---

[13] Part[U] of sth:some but not all of a thing;部分。
[14] Some(det.):used with singular nouns to refer to a person, place, thing or time that is not known or not identified; （与单数名词连用，表示未知或未确指的人、地、事物或时间）某个。
[15] Tip:The world he tromps on is..

terrain[te'rein]*n.*地形

lamppost['læmp‚pəu st]*n.*顶杆;路灯柱

special kind of model called *terrain*. All the buildings, trees, lampposts, and vehicles in our game world are models.

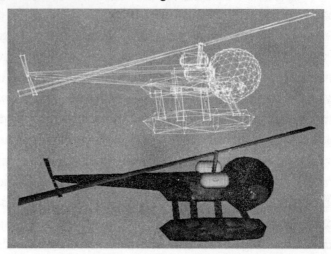

Figure 15.3　A 3D wire-frame and textured models of an old-style helicpter

Textures

In a 3D game, *textures* are an important part of rendering the models in 3D scenes. Textures(in certain cases called *skins*) define the visually rendered appearance of all those models that go into a 3D game. Proper and imaginative uses of textures on 3D models not only will enhance the model's appearance but will also help[16] reduce the complexity of the model. This allow us to draw more models in a given[17] period of time, enhancing performance.

imaginative[i'mædʒi nətiv]*adj.*富于想象的

Sound

Sound provides the contextual flavoring in a 3D game, providing audio cues to events and background sounds that imply environments and context, as well as 3D positioning cues for the player. Judicious use of appropriate *sound effects* is necessary for making a good 3D game.

flavoring['fleivəriŋ]*n.* 调味品;调味香料

cue[kju:]*n.*（戏剧的） 暗示,提示,尾白

judicious[dʒu:'dıʃəs] *adj.*审慎而明智的;有见 地的

Music

Some games, especially *multiplayer games*, use little music. For other games, such as *single-player adventure games*, *music* is an essential tool for establishing *story line moods* and *contextual cues* for the player.

[16] Help—[V **to** inf]。
[17] Given(adj.):already arranged;已经安排好的；规定的。

Composing music for games and pointing out places where music might be use are beyond the scope of this section. However, it[18] is always helpful to pay attention to your game play and whatever mood you are trying to achieve. Adding the right piece of music just might be what you need to achieve the desire mood.

Support Infrastructure

This is more important for persistent multiplayer online games than single player games. When we ponder *game infrastructure* issues, we are considering such things as databases for player scores and capabilities, auto-update tools, Web sites, support forums, and, finally, game administration and player management tools. The following infrastructure items I present them here are to make you aware that you should spend[19] time considering what you might need to do.

ponder['pɔndə]*v.*考虑；琢磨

a. Web Sites

A *Web site* is necessary to provide people with a place to learn news about your game, find links to important or interesting information, and download patches and fixes for your game.

patch[pætʃ]*n.*补丁文件

A Web site provides a focal point for your game, like a storefront. If you intend to sell your game, a well-designed Web site is a necessity.

b. Auto-update

An *auto-update program* accompanies your game onto the player's system. The updater is run at game start-up and connects via the Internet to a site that you specify, looking for updated files, patches, or other data that may have changed since the user last ran the program. It then downloads the appropriate files before launching the game using the updated information.

accompany[ə'kʌmpəni]*v.*伴随

launch[lɔ:ntʃ]*v.*发动；发起（尤指有组织的活动）

Games like Delta Force:Blackhawk Down, World War II Online, and Everquest have an auto-update feature. When you *log in*[20] to the game, the server checks to see if you need to

---

[18] It:used in the position of the subject or object of a verb when the real subject or object is at the end of sentence;用作形式主语或形式宾语，而真正的主语或宾语在句末。

[19] Spend—[VN -ing]。

[20] Log in/on:to perform the actions that allow you to begin using a computer system;登录；注册；进入（计算机系统）。

have[21] any part of your installation upgraded, and if so, it automatically transfers the files to your client. Some auto-updaters will download a local installer program and run it on your machine to ensure that you have the latest files.

c. Support Forums

Community forums or *bulletin boards* are a valuable tool for the developer to provide to customers. Forums are a vibrant community where players can discuss your game, its features, and the matches or games they've played against each other. You can also use forums as a *feedback mechanism*[22] for customer support.

vibrant['vaɪbrənt]*adj.* 充满生机的;生气勃勃的 community[kə'mju:n iti]*n.*社团方式

d. Administrative Tools

If you are developing a persistent online game, it will be important to obtain Web-based tools for creating and deleting player accounts, changing passwords, and managing whatever other uses you might encounter. You will need some sort of hosted Web service with the ability to use CGI-, Perl-, or PHP-based interactive forms or pages. Although this is not strictly necessary, you really should invest in a database to accompany the *administrative tools*.

e. Database

If you intend your game to offer any sort of *persistence* where players' scores, accomplishments, and setting are saved—and need to be protected from fiddling by the players on their own computers—then you probably need a *database back end*. Typically, the administrative tools just mentioned are used to create player records in the *database*, and the game server communicates with the database to authenticate users, fetch and store scores, and save and recall game settings and configurations.

fiddle['fidl]*v.*篡改;伪造

authenticate[ɔ:'θenti ˌkeɪt]*v.* 证明…是真实 的、有效的;证实

A common setup would include MySQL or PostgreSQL or something similar. Again, you will probably need to subscribe to a hosted Web service that offers a database.

---

[21] Have—[VN -ADJ]。
[22] Feedback mechanism:反馈机制。

TERMINOLOGY

3D game	logo
administrative tool	model
auto-update program	mouse cursor
bulletin board	multiplayer game
character	music
community forum	natural physical law
contextual cue	networking
control input	persistence
database	physics formula
database back end	play function
feedback mechanism	scene rendering
First-Person Shooter(FPS)	script
game engine	scripting
game environment	single-player adventure game
game infrastructure	sound
gaming environment	sound effect
graphic	story line mood
graphic environment	terrain
Graphical User Interface(GUI)	texture
GUI bitmap control	textured polygon rendering
GUI button control	Torque Game Engine
GUI text control	upper-left corner
Heads Up Display(HUD)	Web site
log in	

EXERCISES

15.1 Translate each of the following key terms:

    a)3D game

    b)game engine

c)model

d)script

e)Graphical User Interface

f)texture

g)terrain

h)multiplayer game

15.2 Fill in the blanks in each of the following statements:

a)_____ provide most of the essential features of a gaming environment.

b)The architecture of a modern 3D game involves several discrete components:the engine, scripts, GUI, models, textures, _____, and support infrastructure.

c)_____ is one of the most common forms of rendering in FPS games.

d)3D _____ are the essential soul of 3D games.

e)_____ are used to carry the visual appearance of a game and accept users' control inputs.

f)In certain cases, _____ are also called skins.

g)Game support infrastructure refers to auto-update tools, Web sites, support forums, administrative tools, and _____.

h)_____ bring the different parts of a game engine together, provide game play functions, and game world rules.

i)Community forums are also known as _____.

j)The main start-up menus, the dialog boxes, and the various in-game message systems are part of the _____.

15.3 Expand each of the following acronyms:

a)3D.

b)FPS.

c)GUI.

d)HUD.

e)CGI.

f)PHP.

g)SQL.

Reading Material（阅读材料）

Game Environment Optimizations

Modern 3D *gaming platforms* rely heavily on dedicated hardware processing to render images. The goal of the game programmer is generally to maximize throughput through these dedicated processors, and to manage the flow of data between the different components of the machine. Other components of the game machine, such as the CPU and main memory, become less critical as the technology advances.

There are actually two different platforms we consider here. We use the term *video games* to refer to specific, closed systems that connect to a user's TV set. Examples include the Sony Playstation, Microsoft XBOX, Nintendo Gamecube, and so on. *Computer games* run on standard personal computers connected to a *high-refresh-rate* monitor. Each environment poses unique challenges for *Level Of Detail(LOD)*. Game developers design their *engine architecture* around the specific capabilities and limitations of their target platform.

A *video game system* is basically a special-purpose computer designed to efficiently render and display images on a video display. Video game systems are usually less powerful than general computers in terms of their memory, storage, or general processing power, but can be more efficient at tasks such as *3D rasterization*, *3D math*, and *memory manipulation*. The primary design consideration for video game hardware is cost, so components are often highly integrated into just one or two *chips*, and memory(although generally very fast) is often very scarce. Even computer games running on general-purpose hardware, such as a common desktop PC, are moving toward a multiprocessor model that assumes a high-performance graphics co-processor. This *graphics co-processor* performs the majority of the display management, and has its own command set and local memories.

1. Constant Frame Rate

In most genres, a video game is expected to advance at a consistent, fixed *frame rate* of either 30Hz or 60Hz in the NTSC market, or 25Hz or 50Hz in PAL markets. Computer games, on the other hand, have typically been less bound to the *video-standard refresh rates*, and tend to try and achieve the highest possible frame rate up to the maximum refresh rate of the user's monitor, typically between 60 and 100Hz. If a game drops significantly below this refresh rate, the player will experience noticeably "choppy" response times. For commercial video games, the *hardware vendor* has the power to reject games for release if they fail to maintain a minimum acceptable frame rate. Because video game hardware is tried to the television refresh rate, missing the 60Hz performance target by even a small amount will result in an immediate drop to a 30 frames per second, so the costs of failure are quite high.

This requirement for *near-constant refresh* makes level of detail simultaneously more critical and more difficult for a *game environment*. We are forced to avoid methods that require significant periodic recomputation. At the same time, the need to maintain frame rate during different parts of a game requires *real-time load balancing methods* that can adjust LOD on a per-frame basis in response to instantaneous load measurement.

2. Very Low Memory

The biggest challenge facing game developers, particularly on video game platforms, is the relatively low amount of available memory relative to the needs of the game content. This forces developers to be very frugal in the use of memory in all areas, including the use of *compression* when possible. Geometry is generally not the largest user of this precious RAM, but it does require that any level of detail methods used not add considerably to the memory load. LOD methods that can themselves act as a form of compression, such as *parametric terrain representations* or *progressive detail streaming*, are therefore highly prized for games.

Most platforms must also deal with the issue of *segmented memory*, with each dedicated processor having its own *local memory pool*. This means games must not only manage the total amount of memory they use to implement their systems, but this memory must also reside in the correct memory segment, or be moved there efficiently. This becomes a main concern for LOD algorithms that would endeavor to use the main CPU to manipulate geometry information that will be rendered by the *graphics subsystem*, so the cost of moving this memory between systems must be factored into the selection of *LOD management schemes*.

3. Multiple Instantiations

Games typically include multiple *instances* of the same model in a visible *scene*, usually in the form of a number of identical or near-identical "enemies" or "players". The goal of the developer is to share as much data as possible between these instances. When choosing level-of-detail methods, we want to avoid using methods with a *high per-instance memory* or *per-instance computation cost*.

Without LOD, most *game engines* store *mesh* and *model data* in a single shared object that can be referenced multiple times at multiple positions in the game. The basic geometry of the object—including vertex position, normal, color, texture placement, and bone influences—is stored in this shared object. Information unique to one instance (such as its current origin position and orientation) and the positions of any key-framed skeletal bones are stored in the *per-instance object*. The goal is to move as much data as possible into the shared object and instantiate objects with as little memory overhead as possible.

This type of instantiation is typical in many systems and is not unique to games, but games tend to push this memory-saving device further than other applications. Games often use instances not only for the interactive characters that move through the game world, but also for entire rooms, decorative objects, or terrain sections. All this enables the game to present an apparently vast, detailed world within a limited memory budget. Geometry instanced multiple times in a visible scene can also benefit from certain *rendering efficiencies*. If the models in a scene can be sorted according to their shared geometry, they can be tendered sequentially, which may avoid costly *state change*s and, most importantly, *texture cache misses*. On a system with a small high-speed *texture cache*, this can be a critical optimization.

The primary drawback of using instanced geometry in this way is the difficulty in making per-instance modifications of the geometry. For example, a game that wishes to allow the player to deform the geometry at will by adding, removing, or modifying vertex data would be unable to do so in a purely instanced system.

4. Scalable Platforms

Dealing with wide variations in *hardware configurations* was one of the first practical applications of LOD in games. Even before 3D hardware became commonplace, games were expected to run equally well on system with a 90- or 200-MHz processor. This is still a serious concern, but one mainly found on games designed for PC or Macintosh platforms, since dedicated consoles have generally fixed specifications.

With a functional LOD system, the entire platform can be scaled up or down by adding a global bias to the *LOD selection algorithm*. In more extreme cases, we can also add limits to the maximum LOD the game can select under any circumstance. Capping the LOD can also allow us to scale the memory requirements for the game. If the highest-detail model can never be accessed by the game running on a constrained machine, we can avoid even loading the data. *Scaling* on such hardware affects not only geometry but texture resolution selection and the enabling/disabling of various effects.

5. Fill Rate versus Triangle Rate

One major disparity between *video games* and *computer games* is the balance between *fill-rate* and *triangle-rate throughput*. Computer games typically run at *resolution* of 800 ×600 or higher, and often give the user a choice of different screen resolutions. Video games, however, are generally restricted to resolutions of 640×240 or 640×480, depending on whether the game is rendering fields or full frames. In *field rendering*, the game relies on the particulars of the video refresh system, which actually renders a 240-line image twice per frame offset by a single line to give an apparent 480-line image through a process called *interlacing*. Most consoles allow a 640×480 *pixel frame buffer*, which is re-interlaced by the video encoding device for display. As newer video standards begin to take hold, this disparity will begin to disappear. Some current systems support output at video rates above normal *NTSC rates*, such as 480p(640× 480 pixels progressive scanned at 60 Hz) or *HDTV resolutions* such as 720p(720 lines progressive scan) or 1,080i(1,080 lines interlaced).

In either case, a typical video game system has far fewer pixels to rasterize for each frame, up to a *4X advantage* compared to a high-resolution computer game. Because of this disparity, video game applications are more often computationally bound in their ability to transform, light, and set up the rasterization of geometry. In this type of *triangle-rate-limited environment*, geometric LOD methods yield immediate and significant rewards.

6. Average Triangle Size

One metric of particular importance to video game developers is the average triangle size. Typical 3D hardware in the gaming segment is constructed as two separate units:a *geometry-processing unit*(which handles tasks such as triangle setup, lighting, and transformation) and a *pixel-processing unit*(which manipulates the final frame buffer pixels in response to the output from the geometry unit). These are often referred to as the *"front-end"* and *"back-end" units*, respectively. To achieve maximum throughput, these two units need to be carefully load balanced. If the front-end unit is overload, the back-end unit will often be idle, awaiting rasterization jobs from the geometry unit. If the back-end unit is overtaxed, the front-end unit must block, waiting for the rasterizer to accept more commands. Because of the complex interplay between these two parallel pipelines, hardware designers often implement parallel pathways through one or both systems, and provide buffering of data between the two.

The metric of average triangle size helps estimate the balance between the loads on these two systems. A small number of large triangles shift more of the burden to the pixel *back end*, whereas a larger number of tiny triangles will primarily tax the *front end*. Triangle size in terms of screen pixels therefore provides a convenient estimate of this balance. Ideally, an engine would take steps to ensure that the average screen triangle size remains as close as possible to some optimal balance, no matter how the scene objects or observer change over time. Clearly, the other important factor in this relationship will be the cost of the individual triangle or pixel, which is far from fixed. Generally, hardware vendors will advise developers on the ideal triangle size based on different scenarios for vertex and pixel modes. The ideal size for a triangle that uses multiple *dynamic lights* or *skinned blending* would be larger than the size recommendation for a simple unlit triangle, because more pixel operations can be performed between completed triangle computations. Conversely, a more complex pixel operation, would require smaller triangles at equal levels of performance. Clearly some LOD methods are required to maintain our average triangle size near optimum levels.

Note

Lesson 16　Basics of Animation

Once your character is modeled and rigged, you're ready to start[1] animating. *Animation* is a motion-based art, and an understanding of the way[2] objects move is very important to becoming a good *animator*. The laws of motion are the foundation of the science of *physics*, and a little knowledge of physics can go a long way toward[3] giving your characters a sense of realism.

　　Motion is intimately related to time. In fact, motion is simply the change of an object's position over time[4]. *Time* is a *raw material* that actors, comedians, and musicians use constantly. Good comic timing means[5] knowing exactly when to spring the punchline[6]. Good *animation timing* means knowing exactly when your character should react, blink, or pull that huge mallet out from behind his back. *Timing* is the only thing that separates[7] animation from *illustration*. Developing a good sense of timing is very important to becoming a good animator.

　　On top of[8] the basic physics of motion, you also have to consider the meaning of motion. A character moves his body for a reason, and these motions are very important because they convey[9] the character's *mood* and *personality* to the audience. Only through motion can the character truly come to life.[10]

Understanding Motion

If you've ever studied physics, you'll know that motion is the result of forces[11] acting on an object. In order for an object to

rig[rig]v.绑定

realism['ri:əˌlizəm]n.
真实性;逼真
intimately['intimitli]
adv.密切地;紧密地

blink['bliŋk]v.眨眼
mallet['mælit]n.木槌

---

[1] Start—[V -ing]。
[2] Tip:... the way objects move...。
[3] Toward(also towards)(prep):with the aim of obtaining sth, or helping sb to obtain sth;以…为目标（目的）；用于。
[4] Over time:as time passes;随着时间的流逝。
[5] Mean—[V -ing]。
[6] Punchline(noun.)/tag line:the last few words of a joke that make it funny;（笑话最后的）妙趣横生的语句，妙语；画龙点睛之语。
[7] Separate sb/sth from sb/sth:to make sb/sth different in some way from sb/sth else;区分；区别。
[8] On top of sth:in addition to sth;除…之外。
[9] Convey sth to sb:to make ideas, feelings, etc. known to sb;表达，传递（思想、感情等）。
[10] Tip:倒装语句表强调！
[11] 同:...forces that act on an object.

muscle['mʌsl]*n.*肌肉

move or to change direction, a *force* needs to be applied. We all know about the force of gravity, which pulls objects to the ground, but there are plenty of other forces that affect the way characters move, including wind, the weight of a heavy object, or even the forces exerted by a character's own muscles. A strong character moves much differently than a weakling does.

Animation Interfaces

Knowing how an object moves is important for any animator, but another essential kill is being able to create that motion within the computer. All 3D applications are slightly different, but they all use the same basic concepts to define and edit motions. These methods include setting keyframes to define the motion and editing the motion using interfaces such as the motion graph and the dope sheet.

a. Keyframes

specific[spi'sifik]*adj.* 特定的

In the old days of *hand-drawn animation*, animators would draw just the main poses of a character and let *assistant animators* fill in the rest. These main poses are known as *keys* or *keyframes*. In a computerized environment, the· assistant animator is replaced by a computer, and a keyframe is what tells the computer where the object is at a specific time.

b. Motion Graphs

appropriately[ə'prəu pri‚eitli]*adv.*相应地;恰 当地

The frames between keyframes are called, appropriately enough, *inbetweens*. In a computer animation environment, the software calculates your inbetweens, but just as you do in traditional animation, you'll still need to tell the computer exactly how to move the object in between a set of keyframes. The graph itself can be anything from a straight line[12] to a curve. Most motion graphs allow you to adjust the curve using Bézier-type handles, much like the tools used in modeling.

adjust[ə'dʒʌst]*v.*调整; 调节

diagnose['daiəgnəuz] *v.*判断（问题的原因）

Motion graphs are an invaluable tool for the animator in diagnosing and fixing[13] animation problems. Knowing how to read and manipulate motion graphs is an essential skill. Every software package is different, but most motion graphs work in similar ways. Typically, the *horizontal axis* of the graph represents

---

[12] Straight line:直线; curve:曲线。
[13] Fix(verb.):correct sth;校正；解决。

time, while the *vertical axis* represents the parameter being changed, such as position, rotation, scaling, and so on. These parameters are plotted graphically to tell you exactly how an object moves.

plot[plɔt]*v.*（在坐标图上）画出,标出

　　In many cases, the emotion of the object will be broken down into X, Y, and Z components. While this may seem a little complicated at first, using separate[14] curves will ultimately give you more control. You can also use motion graphs to spot problems and tweak a character's motion. Because the motions are graphed out in a smooth curve, any interruption in this curve can signal a glitch or bump in the character's motion.

spot[spɔt]*v.*看出;注意到;发现

tweak[twi:k]*v.*稍稍调整

interruption[ˌintə'rʌpʃən]*n.*中断

glitch[glitʃ]*n.*小差错;小故障

bump[bʌmp]*n.*碰撞;撞击

e. Dope Sheets

　　Another, simpler way to manage keyframes is by using a *dope sheet*. A dope sheet displays just the keyframes without the curves. Typically, the key-frames are displayed as blocks along a linear timeline, with the underlying curves hidden from view. Manipulating the keys changes the timing, but it does not affect the motion graphs; the inbetweens will follow the same curves. Since the keyframes are where you really define the motion, a simplified interface can make[15] editing a scene much easier.

　　Another advantage of dope sheets is that they typically allow for higher-level editing of a character. An animator can select all the keyframes in a character's arm, for example, and reposition them to adjust the entire motion, rather than[16] tweaking it a joint at a time[17].

The Language of Movement

In addition to understanding the forces that affect motion and how to express those forces in a 3D software package, the animator must understand the *language of movement*. Animation has a very specific *vocabulary of motion* that animators can draw from. This vocabulary includes such things as timing, arcs, anticipation, overshoot, secondary action, follow-through, overlap, and moving holds, among others. These

[14]　Separate(adj.):different; not connected;不同的；不相关的。
[15]　Make—[VN -ADJ]。
[16]　Rather than:instead of sb/sth;而不是。
[17]　At a time:separately or in groups of two, three, etc. on each occasion;每次；逐次；依次。

glue[glu:]*n*.胶水

commercial[kəˈmɔːʃəl]
n.（电台或电视播放的）
广告
feature[ˈfiːtʃə]*n*.（电影
的）正片,故事片

verse[vəːs]*n*.歌曲的
段落
chorus[ˈkɔːrəs]*n*.副歌

perceive[pəˈsiːv]*v*. 感
觉到;察觉到;意识到
sequentially[siˈkwen
ʃəli]*adv*.按次序地;从而
recoil[riˈkɔil]*v*.退缩;
畏缩

simultaneously[saim
əlˈteiniəsli]*adv*.同时进
行地;同步地

motions are the raw material; good timing is the glue that holds it all together.

a. Timing

Timing affects every aspect of a film, and on many levels. First, the film is a specific length, anything from a 30-second commercial to a two-hour feature. Second, the cutting of[18] the scenes within this time constraint affects the mood and pace of the film. Third, the acting and timing of the character's actions affect how each individual scene plays.

Think of your film as music. Both film and music rely intimately on time. Your film's scenes can be seen as verses, choruses, or movements. The individual notes of the instruments are the same as the individual actions of your characters. Each action, like each musical note, must be in the right place at the right time. And as with music, bad timing in animation sticks out like a sore thumb.

You must remember that the audience will usually be seeing your film for the first time. As the animator, you need to guide them and tell them exactly where to look at each point in the film. The audience perceives things best sequentially, so you should present your main actions that way, one at a time, within a smooth sequence of motions. A character stubs his toe, recoils, and then reacts. If the reaction is too quick, the audience won't have time to "read" it; the recoil acts as a bridge between the two main actions. The timing between these actions will determine how well the scene reads.

One of the most important lessons you can learn about timing, then, is to draw attention to what is about to move before it moves. An action reads only when the audience is fully focused on it. As the animator, you must guide the audience's eyes through the character's actions.

b. Arcs and Natural Motion

Objects tend to move along *arcs*. This is because an object is usually subject[19] to *multiple forces*, all simultaneously acting

---

[18] Of:used after nouns formed from verbs. The noun after "of" can be either the object or the subject of the action;用于由动词转化的名词之后，of之后的名词可以是受动者，也可以是施动者。

[19] Subject(adj.) to sth:likely to be affected by sth;可能受…影响的。

upon the object. The exceptions are usually *mechanical motions*, which tend to be more linear.

With characters, arcs are also created by the natural mechanics of the body. *Joints* in the body move by rotating, and this rotation creates arcs.

c. Forward Kinematics, Inverse Kinematics[20], and Arcs

When joints are animated using *forward kinematics*, they move by rotating. These rotations automatically move a character's joints along arcs. Such motion is anatomically correct and has a natural look, and it can be produced with little effort.

Limbs animated using *inverse kinematics*, however, need a little more attention. While inverse kinematics is a very helpful tool for locking the legs or arms to a specific location, it works through translation rather than rotation. This means that the software will use translation to move the limb between the keyframes, creating a *straight line* and also an unnatural motion. To correct for this, you may have to add a few extra keyframes or tweak the *motion curves* to get a more natural arc.

d. Force and Drag

When animating, we also need to consider the effects of drag on an object. A force transmitted to an object does not affect all parts of the object equally. Imagine two sticks connected[21] by a flexible joint. If you pulled one of the sticks straight down, the second stick would take a while to "get in line". This effect is called *drag*.

Another point to consider is the way a *multijointed object* will move. If an object has more than two joints, each joint will drag behind the one before it. A third joint added to our stick simply drags behind the second.

The same principles apply to the joints of your character. The spine is really just a collection of joints. Force transmitted to one end of the spine takes time to reach the other end. Force applied to the hand takes time to reach the shoulder and even longer to reach the feet.

exception[ikˈsepʃən]
n. 例外的事物;规则的例外

kinematics[kiniˈmætiks]*n.* 运动学

anatomically[ˌænəˈtɔmikəli]*adv.* 从剖析的角度;从解析的角度

limb[lim]*n.* 肢;臂;腿

flexible[ˈfleksəbl]*adj.* 柔韧的;可弯曲的;有弹性的

---

[20]　Forward kinematics: 正向运动学; inverse kinematics: 反向运动学。
[21]　同:... two sticks that were connected ...

e. Squash and Stretch

Most objects tend to flex and bend as they move. Think of a rubber ball. If you apply a force to the top of it, the ball will *compress* or "*squash*". If you pull the ball from both ends, it will *stretch*.

The fact that objects can change shape when subjected to forces is helpful when animating. You can give [22] a ball bouncing on the ground a "rubbery" feel by squashing it as it hits the ground and stretching it as it takes off.

The principle applies to character animation as well. Characters subjected to forces will change *shape* just as a ball does, but in a more complex way. Think of a gymnast when a character jumps in the air, she'll stretch as she takes off, and when she lands, she'll squash to absorb the force of landing.

f. Anticipation

Anticipation is the body's natural way of gaining momentum before an action beings. When people jump, they swing their arms behind them, bend their knees, and actually move down slightly to get momentum before jumping up. In baseball, a batter will move the bat back before swinging it forward. You lean back before getting out of a chair. It's kind of like getting a head start on the action.

Anticipation is a natural part of motion, and by exaggerating it, we can keep the audience's attention and achieve crisper timing. You'll often need to animate an action that takes place very quickly, such as a "*zip out*". When you do, your audience needs to be fully aware of the action before it occurs, making anticipation of the action very important.

g. Overshoot

As we've seen, anticipation is used to make an action's beginning more lifelike. At the tail end of the action, we have "*overshoot*". In many cases, a character's body will not come to a slow and perfect stop. Instead[23], it will overshoot the stopping point for a few frames and then settle into the pose. Like anticipation, it is a natural part of motion that can be exaggerated to the animator's advantage.

---

gymnast['dʒimˌnæst] *n.*体操运动员

absorb[əb'sɔ:b]*v.*减轻（打击、碰击等的）作用

momentum[məu'mentəm]*n.*冲力

lean[li:n]*v.*前俯

exaggerate[ig'zædʒəreit]*v.*夸张;夸大

crisp[krisp]*adj.*清晰分明的;简明扼要的

perfect['pə:fikt]*adj.*正合适的

---

[22] Give—[VNN]。

[23] Instead(adv.):in the place of sb/sth;代替；反而；却。

When a character throws out his arm to point his finger, his arm will anticipate the move before the action starts. If the motion is quick, the character's arm will naturally overshoot the pose so that the arm is absolutely straight. After a few frames, the arm will then settle into a more natural, relaxed pose. Overshoot can be used to give your character's actions more snap. If you're animating from pose to pose, you can overshoot a pose for a few frames and then settle in.

relaxed[ri'lækst]*adj.*
（人）放松的;自在的

As you've seen, motion can happen for a variety of reasons, but the basics of every motion are the same. All objects move because they are affected by forces, and all moving objects have momentum that makes it difficult to change their course. Additionally, the language of motion defines a number of other specific motions that help bring characters to life, such as anticipation, overshoot, and follow-through.

The best way to understand these concept is to practice. Animate a lot of very simple scenes, using the fundamental motions described above.

TERMINOLOGY

animation	mechanical motion
animation timing	mood
animator	motion
anticipation	motion curve
arc	motion graph
assistant animator	multijointed object
compress	multiple force
dope sheet	overshoot
drag	personality
force	physics
forward kinematics	raw material
hand-drawn animation	shape
horizontal axis	squash

illustration	straight line
inbetween	stretch
inverse kinematics	time
joint	timing
key	vertical axis
keyframe	vocabulary of motion
language of movement	zip out

EXERCISES

16.1 Translate each of the following key terms:

a)animation

b)keyframe

c)force

d)vertical axis

e)timing

f)forward kinematics

g)inbetween

h)mechanical motion

16.2 Fill in the blanks in each of the following statements:

a)_____ is simply the change of an object's position over time.

b)In a computerized environment, a _____ is what tells the computer where the object is at a specific time.

c)_____ is the only thing that separates animation from illustration.

d)_____ are a tool for animators in diagnosing and fixing animation problems by adjusting the cures.

e)A(n) _____ displays just the keyframes as blocks along a linear timeline.

f)Motion is the result of _____ acting on an object.

g)The frames between keyframes are called _____.

h)_____ motions tend to be more linear.

i)Objects tend to move along _____.

j)_____ is the body's natural way of gaming momentum before an action beings.

k)The horizontal axis of a motion graph typically represents _____.

l)Joints in the body move by _____.

m)_____ is used to make an action's tail end more lifelike.

n)When joints are animated using _____ kinematics, they move by rotating.

16.3 Briefly describe each of the following animation features:

a)Keyframes.

b)Motion Graphs.

c)Dope Sheets.

d)Timing.

e)Arcs and Natural Motion.

f)Anticipation.

Reading Material（阅读材料）

What Makes Up a 3D Character?

The primary thing a 3D character is made of would have to be *geometry*. A broad definition of *3D geometry* would be an object that can be edited and rendered. 3ds Max includes a few basics 3D geometric objects called *primitives*. One of the primary primitive objects we use is a box. As a primitive, the ways in which it can be edited are limited to things like the *length*, *height*, *width*, and the *resolution* of each dimension (Figure 16.1). If this were all we could change, it would be extremely difficult to create anything more complicated than box-shaped objects. Instead, we usually adjust the parameters of the primitive box and then convert it into an *editable mesh*, a much more editable form.

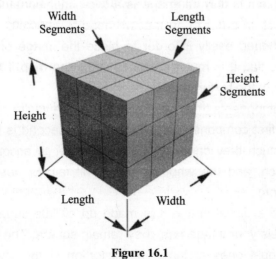

Figure 16.1

This opens up a whole other array of things we can play with. As an editable mesh we now have access to the sub-object parts of the box. These *sub-object parts* are the *bits* and *pieces* that give the box its shape and allow the program to render or draw the surface. The smallest bit is called a vertex. A *vertex* is a point in space denoted by a *plus symbol*(+). Then, between the vertices of an object is what we call an *edge*. It looks simply like a straight line between two vertices, kind of like a three-dimensional connect-the-dots. Then, if we connect three vertices by three edges, we can create a renderable *face*, the third object. If you put a bunch of faces together, you get a larger shape that we call a *mesh*(Figure 16.2).

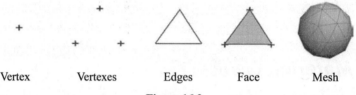

| Vertex | Vertexes | Edges | Face | Mesh |

Figure 16.2

There are several names for meshes; a mesh can be called a *model* or *geometry*. The density of the model or the number of faces in a model defines the *smoothness* of the rendered surface. For example, the digital dinosaurs used in Jurassic Park were made of several million triangles. That was in 1993. Now, the next generation of game engines will be able to handle characters of 5,000 triangles and will look like a model of a million or more. This will give us the ability to create film-quality models for real-time games.

The second most important part of a model is the *texture*. This is the image that will define what the rendered surface will look like. That's easy enough to say, but just a little harder to do. There are two big steps in creating a texture. The first part is called *unwrapping*. The problem is that an image is a two-dimensional thing, and we need to place it on a 3D object. In order to do that we need to separate the model into logical parts that can be flattened easily in order to place the image on it without *distortion*. Because the image is flat, it is much easier to place the map if the model is flat, too. *Stretching* is the bane of the texture artist.

There are two components that come into play when you're creating a *texture map* for a 3D model. The first component is the mesh. The second is the image. They both have limits on how much they can be modified in order to accommodate each other. Change one too much, and the whole process will fail. So let me break down the components a little more.

When you create a digital image it is made up of little squares called *pixels*. No matter how many pixels your image has, they remain square. The geometry is made up of triangles and unequal ones at that. The distortion in the shape of the geometry (triangles) causes the image placed on it to distort, and the pixels appear stretched out. This is because the image is trying to cover a nonsquare distorted surface. An easier way to *imagine* a texture is as a slide in a projector. If you project it onto an uneven surface like a ball or a sheep or a ruffled curtain, the curves of the object distort the image. Texture maps work the same way. Fortunately, you can directly edit the way the map(projectors) sees the surface of the object(screen) by editing the *mapping coordinates*. This is one of the trickier parts of the texturing process.

The second step is to create the image. Before we can do that we need to modify the mapping coordinates until all the geometry is laid out flat. We can take a picture of the flattened geometry and use Photoshop to paint directly on that image. When we apply the image to the geometry, everything should go where we pained it. What to paint and how to do it is a complicated process. There are many different styles and methods for creating images. For the painters out there, you can just paint what you want; for the rest of us, we can use pictures too cut and paste our way to a map. See Figure 16.3.

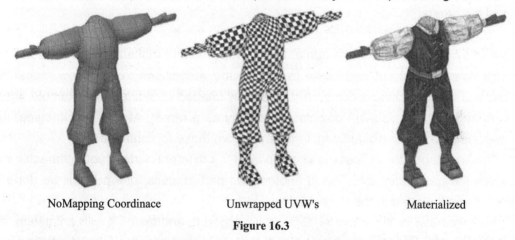

NoMapping Coordinace   Unwrapped UVW's   Materialized

Figure 16.3

The third ingredient is a *skeleton*. I am sure you're asking yourself, "What in the heck do you need a skeleton for?" Well, in the old days if you wanted your model to move around and bend like a human, you needed what was called an *armature*. When animation was done with Play-doh characters, there was a wire that ran inside the "doh" to give it a substructure and support. We will revisit the subject of wire a little bit later. For now, think about your own skeleton. It's the same principle. We have bones to support the muscles that allow us to stand up and move around. They also constrain our movements. Basically, the skeleton acts like your bones, and the mesh simulates the flesh.

I wish I could tell you that this process is completely automated, but that would be a lie. The process of adding a skeleton to your model is called *rigging*, and it has two steps. The first step is to create a skeleton. This is a mostly automated procedure; for example, you just have to click and drag to create a biped humanoid. For creatures other than humanoids, there are easy ways of distorting the bones' shapes, even adding limbs or a tail. Once the skeleton is fitted inside the model we have to connect the two. This is the second step. What we have to do is turn the skeleton into a *deformer* of the mesh and tell the program what bones will control which vertices. This is also mostly an automated process. Unfortunately, we have to check the computer's work on a *bone-by-bone basis*. After all the obvious problems are fixed, we have to do *animation tests* to make sure the rig looks good when the model moves around. This leads us to the second-to-last procedure.

Animation for video games is a little different than you might imagine. *Cyclical animations* or *actions* are strung together to create the illusion of continuous movement from a limited library of motion. So when you press the Jump button, the game runs the jump animation, and when that's done, the game starts the run or stand animation. For every action you see in a game, there has to be an animation that can fit seamlessly with all the others.

Luckily for us, all the *individual animations* for the Unreal Tournament 2004 characters are included with the game. So we don't have to make all the animation from scratch. We will just cut and paste to add all the animations to your new character, unless you want to make a completely differently shaped character. For example, there is a raptor pack that you can download and play as a raptor. As you can imagine, the animations for a raptor would be quite different from those for a human.

The last step in the process is to *compile* all the different parts of your character into *files* that the game can use. This is another two-part process; one part will be done in 3ds Max and the other in the Unreal Editor.

After we have built our model, textural it , rigged it, and tested it with animation, it's ready to go. We will use a plug-in called *Actor-X* to export your models' information in a form that can be understood by the Unreal Editor. Then, from the Unreal Editor, there will be one more complication into files that the *game engine* can understand. It sounds complicated, but this is one of the easiest parts of the whole process and should take about an hour. Then we can play our new character in the game.

Note

Appendix A   Key to Verb Patterns

A.

Intransitive verbs 不及物动词

    [V]　verb used alone 单独使用的动词

        *A large dog **appeared**.*

    [V+adv./prep.]　verb+adverb or prepositional phrase 动词+副词或介词短语

        *A group of swans **floated by**.*

B.

Transitive verbs 及物动词

    [VN]　verb+noun phrase 动词+名词短语

        *Jill's behaviour **annoyed me**.*

    [VN+adv./prep.]　verb+noun phrase+noun phrase or prepositional phrase

                动词+名词短语+副词或介词短语

        *He **kicked the ball into** the net.*

C.

Transitive verbs with two objects 后接双宾语的及物动词

    [VNN]　verb+noun phrase+noun phrase 动词+名词短语+名词短语

        *I **gave Sue the book**.*

D.

Linking verbs 连系动词

    [V-ADJ]　verb+adjective 动词+形容词

        *His voice **sounds hoarse**.*

    [V-N]　verb+noun phrase 动词+名词短语

        *Elena **became a doctor**.*

    [VN-ADJ]　verb+noun phrase+adjective 动词+名词短语+形容词

        *She **considered herself lucky**.*

    [VN-N]　verb+noun phrase+noun phrase 动词+名词短语+名词短语

        *They **elected him president**.*

E.

Verbs used with clauses or phrases 后接从句或短语的动词

    [V **that**] [V (**that**)]　verb+**that** clause 动词+that 从句

        *He **said that** he would walk.*

    [VN **that**] [VN (**that**)]　verb+noun phrase+**that** clause 动词+名词短语+that 从句

        *Can you **remind me that** I need to buy some milk?*

[V **wh-**]   verb+**wh**-clause 动词+**wh**-从句

*I **wonder what** the job will be like.*

[VN **wh-**]   verb+noun phrase+**wh**-clause 动词+名词短语+wh-从句

*I **asked him where** the hall was.*

[V **to**]   verb+**to** infinitive 动词+带 to 的不定式

*I want **to leave** now.*

[VN **to**]   verb+noun phrase+**to** infinitive 动词+名词短语+带 to 的不定式

*I **forced him to go** with me.*

[VN inf]   verb+noun phrase+infinitive without "to" 动词+名词短语+不带 to 的不定式

*Did you **hear the phone ring**?*

[V **-ing**]   verb+**-ing** phrase 动词+**-ing** 短语

*She never **stops talking**.*

[VN **-ing**]   verb+noun phrase+**-ing** phrase 动词+名词短语+-ing 短语

*His comments **set me thinking**.*

F.

Verbs+direct speech 动词+直接引语

[V **speech**]   verb+direct speech 动词+直接引语

*"It's snowing," she **said**.*

[VN **speech**]   verb+direct speech 动词+直接引语

*"Tom's coming too," she **told him**.*

Appendix B   Relative Clauses and Other Types of Clause

B.1   Relative Pronouns

A.

Defining and *non-defining relative clauses* begin with a *relative pronoun*, which can sometimes be omitted:

❑   We went to a beach (**which/that**) Jane had recommended to us.

Here the relative pronoun refers to "a beach", and the subject of the relative clause is "Jane". Compare:

❑   I know a man **who/that** ran in the New York Marathon last year.

Where the relative pronoun refers to "a man", and the subject of the relative clause is also "a man". In this case, the relative pronoun can't be omitted.

B.

When we use *a defining relative clause*, the relative pronoun can be either the subject or the object of the relative clause. When it is the *subject* the word order is subject + verb + object:

❑   I have *a friend* **who/that** *plays guitar*. (a friend = subject, plays = verb, guitar = object)

When the relative pronoun is the *object* the word order is object + subject + verb:

❑   He showed me *the rocks* (**which/that**) *he had collected*. (the rocks = object, he = subject, had collected = verb)

C.

Relative pronouns are used to add information in *defining relative clauses* as follows:

adding information about things

Relative pronoun	which	that	no relative pronoun
subject	√	√	×
object	√	√	√

adding information about people

Relative pronoun	which	that	no relative pronoun	whom
subject	√	√	×	×
object	√	√	√	√

✧   When we add information about things, we can use **that** (or **no relative pronon**) as object in conversation and **which** in more formal contexts:

❑   Decorating's a job (**that**) I hate. (*rather than* "...which... " in this informal context)

✧   When we add information about people, we generally prefer **that** (or **no**

relative pronoun) as object in informal contexts rather than **who** or **whom**:

- ☐ That's the man (**that**) I met at Alison's party. (rather than ...who/whom I met...)
- ✧ **whom** is very formal and rarely used in spoken English:
- ☐ The boy **whom** Elena had shouted at smiled. (*less formally* **that**, **no relative pronoun** or **who**)
- ✧ We use **that** as subject after:**something** and **anything**; words such as **all**, **little**, **much**, and none used as pronouns; and noun phrases that include superlatives. **Which** is also used as subject after **something** and **anything**, but less commonly:
- ☐ These walls are *all* **that** *remain* of the city. (*not* ...which remain of the city.)
- ✧ Note that we can use **that** (or **no relative pronoun**) as object after **something/ anything**; **all**, etc.; and noun phrases with superlatives. For example:
- ☐ She's one of the kindest people (**that**) I know. (*not* ...one of the kindest people who I know.)

D.

Relative pronouns are used to add information in *non-defining relative clauses* as follows:

adding information about things

Relative pronoun	which	that
subject	√	√
object	√	√

adding information about people

Relative pronoun	who	whom
subject	√	×
object	√	√

- ✧ Notice that we must include a relative pronoun in a non-defining relative clause.
- ✧ We can use **who** or **whom** as object, although **whom** is very formal:
- ☐ Professor Johnson, **who(m)** I have long admired, is to visit the University next week.
- ✧ When we add information about things, we can use **which** as subject or object. **That** is sometimes used instead of **which**, but some people think this is incorrect:
- ☐ The Master's course, **which** I took in 1990, is no longer taught. (*or* ...**that** I took...)

B.2   Other relative words:whose, when, whereby, etc.

A. Clauses with whose

We use a relative clause beginning with the relative pronoun **whose + noun**, particularly in written English, when we talk about we talk about something belonging to or associated with a person, animal or plant:

- ☐ Stevenson is an architect **whose designs** have won international praise.

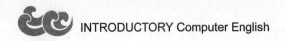

- ☐ Sue was taking care of a rabbit **whose ears** were badly damaged in a fight with a cat.

We can use **whose** in both *defining* and *non-defining relative clauses*.

We generally avoid using **whose** to talk about something belonging to or associated with a *thing*:

- ☐ I received a letter, and its poor spelling made me think it was written by a child. (*more natural than* "I received a letter, **whose** poor spelling made me think... ")

However, we sometimes use **whose** when we talk about towns, countries, or organizations:

- ☐ The film was made in *Botswana*, **whose wildlife parks** are lager than those in Kenya.
- ☐ We need to learn from *companies* **whose trading** is more healthy than our own.

In academic writing **whose** is used to talk about a wide variety of "belonging to" relationships:

- ☐ Students are encouraged to use an appropriate theory in order to solve *problems* **whose** geographical limits are clear.

B. Clauses with when, whereby, where and why

We can begin relative and other clauses with **when** (referring to time), **whereby**(method or means; used mainly in formal contexts), and **where**(loaction). In formal English in particular, a phrase with **preposition + which** can often be used instead of these:

- ☐ He wasn't looking forward to the time **when** he would have to leave. (*or* ...the time **at which**...)
- ☐ Do you know the date **when** we have to hand in the essay? (*or* ...the date **on/by which**...)
- ☐ The government is to end the system **whereby** (= "by which means") farmers make more money from leaving land unplanted than from growing wheat. (*or* ...the system **in/by which** farmers...)
- ☐ This was the place **where** we first met. (*or* ...the place **at/in which** we...)

In academic English, we can also use **where** to refer to relationships other then location, particularly after words such as **case**, **condition**, **example**, **situation**, **system**:

- ☐ Later in this chapter we will introduce *cases* **where** consumer complaints have resulted in changes in the law. (*or more formally* ...cases **in which**...)

We can also use **a/the reason why** or **a/the reason that** or just **a/the reason**:

- ☐ I didn't get a pay rise, but this wasn't **the reason why** I left. (*or* ...**the reason (that)** I left.)

C. Clauses with who and what; whatever, whoever and whichever

Some clauses beginning with a *wh*-word are used like a noun phrase in a sentence. There are sometimes called *nominal relative clauses*:

- □ Can you give me a list of **who's** been invited? (= the people who have been invited)
- □ I didn't know **what** I should do next. (= the thing that I should do next)

Notice that we can't use **what** in this way after a noun:

- □ I managed to get all the *books* **that** you asked for. (*not* ...all the books what you asked for.)

We use clauses beginning with **whatever** (= anything *or* it doesn't matter what), **whoever** (= the person/group who *or* any person/group who), or **whichever** (= one thing or person from a limited number, to talk about things or people that are indefinite or unknown):

- □ I'm sure I'll enjoy **whatever** you cook.
- □ **Whoever** wins will go on to play Barcelona in the final.
- □ **Whichever** one of you broke the window will have to pay for it.

B.3   Prepositions in relative clauses

A.

In formal styles **noun + of which** is often preferred to **whose + noun** when we talk about things:

- □ A huge amount of oil was spilled, *the effects* **of which** are still being felt.
- □ The end of the war, *the anniversary* **of which** is on the 16th November, will be commemorated in cities throughout the country.

We can use **of which** and **of whose**, but not usually **which** or **whose**, after **all**, **both**, **each**, **many**, **most**, **neither**, **none**, **part**, **some**, **a number** (one, two, etc.; the first, the second, etc.; half, a third, etc.) and **superlatives** (the best, the biggest, etc.):

- □ Lotta was able to switch between German and Russian, **both of which** she spoke fluently. (*not* ...both which she spoke fluently.)
- □ She jointed the local tennis club, **most of whose** members were at least 60.

In formal contexts, **of which** can be used instead of **that/which...of** in relative clauses:

- □ The school **that/which** she is head of is closing. (*or more formally* The school **of which** she...)
- □ The book **that/which** he's most proud of... (*or more formally* The book **of which** he...)

B.

In formal, mainly written, English **whose** can come after a preposition in a relative clause. However, it is more natural to put the preposition at the end of the clause in less formal contexts and in spoken English:

- ❑ The council is in discussion with Lord Thomas, **on whose** land most of the village is built. (*or less formally* ...Lord Thomas, **whose** land most of the village is built **on**.)
- ❑ I now turn to Fred, **from whose** work the following quotation is taken. (*or less formally* ...Freud, **whose** work the following quotation is taken **from**.)

C.

When a preposition is needed with the relative pronoun **which** and **whom** we usually put it before the relative pronoun in formal styles:

- ❑ The rate **at which** a material heats up depends on its chemical composition.
- ❑ Her many friends, **among whom** I like to be considered, gave her encouragement.

After a preposition we usually use **whom** rather than **who** in formal styles:

- ❑ Is it right that politicians should make important decisions without consulting the public **to whom** they are accountable? (*rather than* ...the public to who they are accountable.)

and we don't use **that** or **no relative pronoun**:

- ❑ The valley **in which** the town lies is heavily polluted. (*not* The valley in that the town lies is heavily polluted.; *not* The valley in the town lies is heavily polluted.)

In less formal English we usually put the preposition later in the relative clauses clause rather than at the beginning:

- ❑ The office **that** Graham took us **to** was filled with books.

and we prefer **who** (or **that**) rather than **whom**:

- ❑ The playground wasn't used by the children **who** it was built **for**.

D.

If the verb in the relative clause is a two-word verb (e.g. **come across**, **fill in**, **look after**, **take on**), we don't usually put the preposition before the relative pronoun:

The Roman coins, **which** a local farmer **came across** in a filed, are now on display in the National Museum. (*not* ...coins, across which the local farmer came, are...)

With three-word verbs, we only put the preposition before the relative pronoun in a very formal or literary style, and many people avoid this pattern:

She is one of the few people **to whom I took up**. (*or less formally* ...**who I took up to**.)

B.4   Other ways of adding information to noun phrases

A.

We sometimes add information about a person or thing referred to in one noun phrase by talking about the same person or thing in a different way in a following noun phrase:

- *A hooded cobra, one of the world's most dangerous snakes*, has escaped from Dudley Zoo.
- *Dr Alex Parr, director of the State Museum,* is to become the government's arts adviser.
- When Tom fell off his bike we gave him *arnica, a medicine made from a flower,* for the bruising.

In writing, the items are usually separated by a comma, and in speech they are often separated by a pause or other intonation break. However, when the second item acts like a defining relative clause, when it is usually a name, there is usually no punctuation in writing or intonation break in speech:

- *My friend Jim* has moved to Sweden. (*rather than* My friend, Jim, ...)
- The current champion is expected to survive her first-round match with *the Italian Silvia Farina*. (*rather than* ...the Italian, Silvia Farina.)

B.

We can add information to a noun phrase with a conjunction such as **and** or **or**:

- Kurt Svensson, her teacher **and** *well-known concert pianist*, thinks that she has great talent. (= her teacher is well-known concert pianist)
- My business partner **and** *great friend* Tom Edwards is getting married today.
- Phonetics **or** *the study of speech sounds* is a common component on courses in teaching English as a foreign language.

C.

The adverb **namely** and the phrase **that is** are used to add details about a noun phrase:

- This side-effect of the treatment, **namely** *weight gain*, is counteracted with other drugs.
- The main cause of global warming, **that is** *the buring of fossil fuels*, is to be the fucus of negotiations at the international conference.

D.

We can also add information to a noun phrase using a participle clause beginning with an **-ing**, **-ed** or **being + -ed** verb form. These are often similar to *defining relative clauses*:

- Any passengers *travelling to Cambridge* should sit in the first two carriages of the train. (*or* Any passengers who are travelling...)
- The people *living next door* come from Italy. (*or* The people who are living next door...)
- The weapon *used in the murder* has now been found. (*or* The weapon that was used...)
- The book *published last week* is his first novel. (*or* The book that was published last week...)

- ❏ The prisoners **being released** are all women. (*or* The prisoners who are being released...)
- ❏ The boys **being chosen** *for the team* are under 9. (*or* The boys who are being chosen...)

Notice that **-ing** participle clauses correspond to defining relative clauses with an active verb, while **-ed** and **being+-ed** clauses correspond to defining relative clauses with a passive verb.

We can also use a **to-infinitive clause**, as in:

- ❏ Have you brought a book **to read**? (= you bring it and you read it)
- ❏ Have you brought a book for Kevin **to read**? (= you bring it and Kevin reads it)
- ❏ My decision **to resign** *from the company* was made after a great deal of thought.
- ❏ I thought that the decision of the committee, **to increase** *staff holidays*, was a good one.

E.

In written English, particularly in newspapers, **-ing** and **-ed** clauses are also used instead of *non-defining relative clauses*. There are usually written between commas or dashes (-):

- ❏ The men, *wearing anoraks and hats*, made off in a stolen Volvo estate.
- ❏ The proposals-*expected to be agreed by ministers* - are less radical than many employers had feared.

F.

We commonly add information about a thing or person using a prepositional phrase. Often these have a meaning similar to a relative clause:

- ❏ What's the name of the *man* **by** the window? (*or* ...the man **who's** by the window?)
- ❏ It's in the *cupboard* **under** the stairs. (*or* ...the cupboard **that's** under the stairs.)
- ❏ She lives in the *house* **with** the red door. (*or* ...the house **which has** the red door.)

In some cases, however, these prepositional phrases do not have a corresponding relative clauses:

- ❏ You need to keep a careful *record* **of** what you spend.
- ❏ There is likely to be an *increase* **in** temperature tomorrow.

We often prefer a relative clause rather than a prepositional phrase in non-defining relative clauses with **be + preposition** or with **have** as a main verb:

- ❏ *Johnson*, **who was** in the store at the time of the robbery, was able to identify two of the men. (*rather than* ...Johnson, in the store...)

□   *Jim Morton*, **who has** a farm in Devon, has decided to grow only organic vegetables. (*rather than* Jim Morton, with a farm in Devon, has...)

G.

In written English, particularly in academic writing, a series of prepositional phrases and relative clauses is often used to add information about a previous noun phrase. Note that prepositonal phrases can be used with an adverbial function (e.g. "...taken the drug *in the last 6 months*" in the sentence below):

□   Doctors are contacting patients with diabetes who have taken the drug in the last 6 months.

□   Scientists in Spain who have developed the technique are optimistic that it will be widely used in laboratories within the next decade.

We can also use participle clauses and noun phrases in a series of clauses/phrases which add information add information to the preceding noun phrase:

□   The waxwing is the only bird found in Britain with yellow and red tail features.

□   Mr Bob Timms, leader of the Democratic Party, MP for Threeoaks, has announced his resignation.

H.

Notice that adding a series of prepositional phrases can often lead to ambiguity. For example:

□   The protesters were demonstrating against the mistreatment of animals on farms.

Could mean either that the place the protesters were demonstrating was "on farms" or that the animals were "on farms". We could make the sentence unambiguous with, for example:

□   The protesters were demonstrating on farms against the mistreatment of animals. *or*

□   The protesters were demonstrating against the mistreatment of animals kept on farms.

B.5   Participle clauses with adverbial meaning

A.

We can use **present participle (-ing)** and **past participle (-ed)** clauses with an adverbial meaning. Clauses like these often give information about the timing, causes, and results of the events described:

□   *Opening her eyes*, the baby began to cry. (= When she opened her eyes...)

□   *Faced with a billed for £10.000*, John has taken an extra job. (= Because he is faced...)

□   *Looked after carefully*, the plant can live through the winter. (= If it is looked after...)

- *Having completed the book*, he had a holiday. (perfect; = When/Because he had completed...)
- The first was expensive, *being imported*. (simple passive; = because it was imported)
- *Having been hunted close to extinction*, the rhino is once again common in this area. (perfect passive; = Although it had been hunted close to extinction...)

B.

The implied subject of a participle clause (that is, a subject known but not directly mentioned) is usually the same as the subject of the main clause:

- *Arriving* at the party, we saw Ruth standing alone. (= When **we** arrived...**we** saw...)

However, sometimes the implied subject is not referred to in the main clause:

- *Having wanted* to drive a train all his life, this was an opportunity not to be missed.

In careful speech and writing we avoid different subjects for the participle and main clause:

- Turning round quickly, the door hit me in the face. (first implied subject = "I"; second subject = "the door")

C.

In formal English, the participle clause sometimes has its own subject, which is often a pronoun or includes one;

- The collection of vases is priceless, **some** *being over two thousand year old*.
- **Her voice** *breaking with emotion*, Jean spoken about her father's illness.

We use a present participle (**-ing**) clause to talk about something happening at the same time as an event in the main clause, or to give information about the facts given in the main clause.

D.

When we use **not** in a participle clause it usually comes before the participle. However, it can follow the participle, depending on the part of the sentences affected by **not**. Compare:

- *Wishing* **not** to go out that night, I made an excuse. ("not" relates to "to go out that night"; the sentence means "I didn't want to go out on that particular night") *and*
- **Not** *wishing* to go out that night, I made an excuse. ("not" relates to "wish to go out that night"; the sentence could mean "going out on that paticular night wasn't my wish")

E.

We use a clause beginning with **having + past participle** rather than a present participle if the action in the main clause is the consequence of the event in the participle clause:

- ☐ **Having won** every major judo title, Mark retired from international competition. (*or* **After winning**...; *not* Winning every major judo title...)
- ☐ **Having broken** her leg the last time she went, Brenda decided not to go on the school skiing trip this year. (*or* **After breaking** her leg...; *not* Breaking her leg...)

We can use either a **present participle (-ing)** clause or a **having + past participle** clause with a similar meaning when the action in the participle clause is complete before the action in the main clause begins. Compare:

- ☐ **Taking off** his shoes, Ray walked into the house. (*Having taken off*...has a similar meaning) *and*
- ☐ **Running** across the field, I fell and hurt my ankle. (= While I was running...; "Having run..." would suggest that I fell *after* I had run across the field)

F.

We can use prepositions such as **after**, **before**, **besides**, **by**, **in**, **on**, **since**, **through**, **while**, **with**, and **without** with a present participle (**-ing**) clause with an adverbial meaning:

- ☐ **While understanding** her problem, I don't know how I can help. (= Although I understand...)
- ☐ **After spending** so much money on the car, I can't afford a holiday.
- ☐ **Before being changed** last year, the speed limit was 70 kph. (passive form)

Less formal alternatives have a clause with a verb that can change according to tense and subject. Compare:

- ☐ **Since moving** to London, we haven't had time to go to the theatre. *and*
- ☐ **Since we moved** to London, we haven't had time to go to the theatre. (less formal)

G. by, in, on + ing

☐ **By working** hard, she passed her maths exam. ☐ They only survived **by eating** roots and berries in the forest.	= the **-ing** clause indicates "the method or means used"
☐ **On returning** from Beijing, he wrote to the Chinese embassy. ☐ John was the first person I saw **on leaving** hospital.	= the **-ing** clause indicates "when"
☐ **In criticising** the painting, I knew I would offend her. ☐ **In choosing** Marco, the People's Party has moved to the left.	= the **-ing** clause indicates "cause"

We can often use **by + -ing** or **in + -ing** with a similar meaning, although **by + -ing** is preferred in informal contexts:

- **In/By writing** the essay about Spanish culture, I came to understand the country better. ("In writing..." = the consequence of writing was to understand...; "By writing..." = the method I used to understand the country better was to write...)

But compare:

- **By telephoning** every hour, she managed to speak to the doctor. (*not* In telephoning...; the method, not the consequence)

H. with -ing; without -ing

With + -ing often introduces a reason for something in the main clause. This use is fairly informal. Notice that a subject has to come between **with** and **-ing**:

- **With** Louise **living** in Spain, we don't see her often. (= Because Louise lives in Spain...)
- **With** sunshine **streaming** through the window, Hugh found it impossible to sleep. (= Because the sunshine was streaming...)

With and **what with** can also be used with a noun phrase to introduce a reason:

- **With** *my bad back* I won't be able to lift a heavy suitcase.
- **What with** *the traffic* and *the heavy rain*, it's no wonder you were late.

We can use **without + -ing** to say that a second action doesn't happen:

- I went to work **without eating** breakfast.
- They left without paying.

Often, however, it has a similar meaning to "although...not" or "unless":

- **Without meaning** to, I seem to have offended her. (= Although I didn't mean to...)
- **Without seeing** the photo, I can't judge how good it is. (= Unless I see the photo...)

I.

Adverbial meanings can also be added by a clause beginning with a conjunction or adjective but with no verb, having the same meaning as a clause beginning with a **conjunction + subject + be**. This is used in fairly formal English. More informal alternatives are given in brackets:

- **While in Poland**, they will play two concerts in Warsaw. (*or* **While** they are in Poland...)
- **Although just two feet apart**, they didn't speak. (*or* **Although** they were just...)

☐ I try to use public transport **whenever possible**. (*or* ...**whenever** it is possible.)

☐ **Unhappy with the decision**, Johnson swore at the referee. (*or* **Because** he was unhappy...)

☐ James relaxed, **pleased with his day's work**. (*or*...**because** he was pleased...)

Vocabulary Index

L1

device[di'vais]*n.*装置;设备;部件
logical['lɔdʒikəl]*adj.*符合逻辑的;按照逻辑的
ponder['pɔndə]*v.*沉思;考虑;琢磨
function['fʌŋkʃən]*v.*运转;工作
process['prəuses]*v.*数据处理
instruction[in'strʌkʃən]*n.*（计算机的）指令
orderly['ɔ:dəli]*adj.*有条理的
envision[en'viʒən]*v.*想象;设想
facilitate[fə'siliteit]*v.*促进;促使;使便利
audio['ɔ:djəu]*n.*（指方法）录音
video['vidiəu]*n.*（指方法）录像,录影
access['ækses]*n.*访问,存取（计算机文件）
temporary['tempərəri]*adj.*临时的;暂时的
available[ə'veiləbl]*adj.*可获得的;可找到的
volatile['vɔlətail]*adj.*不稳定的;可能急剧波动的
erase[i'reiz]*v.*抹去;清洗（磁带上的录音或存储器中的信息）
arithmetic[ə'riθmətik]*n.*算术运算,四则运算
mechanism['mekənizəm]*n.*方法;机制
coordinator[kəu'ɔ:dineitə]*n.*协调器,协调者
supervise['sju:pəvaiz]*v.*监督;管理
retrieve[ri'tri:v]*v.*检索数据
interact[ˌintər'ækt]*v.*相互作用
address[ə'dres]*v.*处理;应对;设法解决
costs[kɔsts]*n.*成本
dramatically[drə'mætikli]*adv.*（变化、事情等）突然地;巨大地
refinement[ri'fainmənt]*n.*精炼;提炼

L2

encode[en'kəud]v.把…编码
numerical/numeric[nju:'merikəl]adj.数字的;用数字表示的
repercussion[ˌri:pə'kʌʃən]n.（间接的）影响;反响;恶果
successive[sək'sesiv]adj.连续的;连接的;相继的
proliferation[prəuˌlifə'reiʃən]n.激增;涌现
alleviate[ə'li:vieit]n.减轻;缓解
dominant['dɔminənt]adj.占支配地位的;占优势的;显著的
manufacturer[ˌmænju'fæktʃərə]n.生产商;制造者
elaborate[i'læbəreit]adj.复杂的;详尽的
numerous['nju:mərəs]adj.很多的;许多的
proprietary[prə'praiəˌteri:]adj.专有的;所有的
shortly['ʃɔ:tli]adv 立刻;马上
notation[nəu'teiʃən]n.（数学、科学和音乐中用于表示信息的）符号;记号
fractional['frækʃənəl]adj.分数的;小数的
classify['klæsifai]v.将…分类;将…归类
dot[dɔt]n.点;小圆点
facsimile=fax[fæk'siməli:]n.传真机
intensity[in'tensiti]n.强度
chrominance['krəuminəns]n.色度（任一颜色与亮度相同的指定参考色之间的色差）
rescale[ri:'skeil]v.重新调节;重新缩放
arbitrary['ɑ:bitrəri]adj.任意的;随心所欲的
grainy['greini]adj.粒状的;有颗粒的
sample['sɑ:mpl]v.取样;采样
amplitude['æmpliˌru:d]n.（声音、无线电波等的）振幅
fidelity[fi'deliti]n.高精度;高准确性
stereo['stiəriəu]n.立体声
synthesizer['sinθisaizə]n.语音合成器

L3

roughly['rʌfli]adv.粗略地;大体上;大致上
accordingly[ə'kɔ:diŋli]adv.（尤用于句首）因此;于是
coordinate[kəu'ɔ:dineit]v.使…得以协调

specific[spi'sifik]*adj*.（名词前）特定的
conflicting[kən'fliktiŋ]*adj*.相互矛盾的;相冲突的
request[ri'kwest]*n*.（计算机的）请求
fairly['fɛəli]*adv*.公平地;公正地
carious['kɛəriəs]*adj*.（力量、影响等）衰弱,衰败;精疲力尽的
handheld['hænd,held]*adj*.手持的
universally[,ju:ni'və:səli]*adv*.全体地;一致地
vendor['vendə]*n*.销售商
ship[ʃip]*v*.上市;把…推向市场
lack[læk]*v*.缺少
contradictory[,kɔntrə'diktəri:]*adj*.相互矛盾的;对立的
concentrate['kɔnsəntreit]*v*.集中（注意力）;聚精会神
optimal['ɔptəməl]*adj*.最佳的;最优的
facilitate[fə'siliteit]*v*.（正式地）促进;使便利

L4

objective[əb'dʒektiv]*n*.目标;目的
solving['sɔlviŋ]*n*.解决;处理;求解
illustrate['iləstreit]*v*.（用示例、图画等）说明;解释
depict[di'pikt]*v*.描写;描述
participate[pɑ:'tisipeit]*v*.（正式地）参加;参与
existing[ig'zistiŋ]*adj*.现存的
stakeholder['steikhəuldə]*n*.（某组织、工程、体系等的）参与者,参与方;有权益关系着
feasible['fi:zəbl]*adj*.可行的;行得通的
address[ə'dres]*v*.设法解决;处理
outweigh[aut'wei]*v*.大于;超过
framework['freimwə:k]*n*.（体系的）结构,机制;架构
pose[pəuz]*v*.造成（问题、威胁等）;引起
fulfill[ful'fil]*v*.履行;实现
overall['əuvərɔ:l]*adj*.全面的;总体的
deliver[di'livə]*v*.清楚表述
remaining[ri'meiniŋ]*adj*.其余的
portion['pɔ:ʃən]*n*.部分
exhaustive[ig'zɔ:stiv]*adj*.详尽的;彻底的;全面的

续表

thorough['θʌrə]*adj.*深入的;细致的
ascertain[ˌæsə'tein]*v.*弄清;查明
underrate[ˌʌndə'reit]*v.*低估;过低评价
budget['bʌdʒit]*n.*预算
glossing[glɔ:s]*n.*虚假外表;虚饰
devise[di'vaiz]*v.*设计
systematic[ˌsistə'mætik]*adj.*系统化的;成体系的;条理化的
schedule['ʃedju:əl]*n.*日程安排;工作计划
verify['verifai]*v.*核实;查对

L5

overall['əuvərɔ:l]*adj.*全面的;总体的
mandatory['mændəˌtɔ:ri]*adj.*强制的
manipulate[mə'nipjuleit]*v.*（熟练地）操作,使用
essential[i'senʃəl]*n.*要点;要素
attempt[ə'tempt]*n.*尝试;试图
reference['refrəns]*v.*查阅;索引
analogy[ə'nælədʒi]*n.*类比;比喻
hierarchical[ˌhaiə'ra:kikəl]*adj.*层次性的
inheritance[in'heritəns]*n.*继承
invocation[ˌinvə'keiʃən]*n.*调用
syntactical[sin'tæktikəl]*adj.*句法的
finite['fainait]*adj.*有限的;有限制的
denote[di'nəut]*v.*表示;标识
incident['insidənt]*adj.*伴随而来的;自然的
adjacent[ə'dʒeisənt]*adj.*邻接的;毗邻的

L6

flowchart['fləuˌtʃa:t]*n.*流程图
notably['nəutəbli]*adv.*尤其;特别
parameter[pə'ræmitə]*n.*参数
demonstrate['demənstreit]*v.*示范;演示
occurrence[ə'kʌrəns]*n.*事情

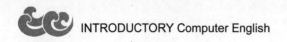

modify['mɔdifait]v.调整;稍作修改
approach[ə'prəutʃ]n.方式;方法
oversimplify['əuvə'simplifai]v.陈述过于简略;说明过于简化
conceptually[kən'septjuəli]adv.在概念上
initial[i'niʃəl]adj.最初的;开始的;初期的
determine[di'tə:min]v.确定;裁决
instantiate[in'stænʃieit]v.实例化
distinction[dis'tiŋkʃən]n.差别;区分
varying['vɛəriŋ]adj.变化着的
encapsulate[en'kæpsə,leit]v.封装

L7

robust[rəu'bʌst]adj.强壮的;强劲的
rover['rəuvə]n.漫游者
hype[haip]adj.言过其实的;大肆宣传的
attractive[ə'træktiv]adj.（事物）有吸引力的;吸引人的
versatile['və:sətail]adj.多功能的;多用途的
buzzword['bʌz,wə:d]n.（出版物上的）时髦用语;流行行话
model['mɔdəl]v.复制;仿制
construct[kən'strʌkt]n.构件;结构体
eliminate[i'limineit]v.排除;去除
syntax['sin,tæks]n.（计算机上的）句法;语构
negative['negətiv]adj.消极的;负面的
inherently[in'hiərəntli]adv.天生地;固有地
paradigm['pærə,daim]n.范例;样式
modularity[,mɔdju'læriti]n.模块化
elitist[ei'li:tist]n.杰出物;精品
detect[di'tekt]v.发现;查出
corrupt[kə'rʌpt]v.引起（计算机文件、数据等的）错误;破坏
exception[ik'sepʃən]n.异常;例外的事物
terminate['tə:mineit]v.结束;终结
harm[hɑ:m]n.危害;损害
stray[strei]adj.不守规矩的;偏离的
premise['premis]n.前提;假定

neutral['nju:trəl]*adj.*中立的;无倾向性的
stem[stem]*v.*抑制;遏制
arithmetic[ə'riθmətik]*n.*算术;计算
simultaneously[saiməl'teiniəsli]*adv.*同时进行地;同步地
transparently[træns'pɛərənt]*adv.*透明地;方便易懂地

L8

scene[si:n]*n.*情景;场景
corporation[ˌkɔ:pə'reiʃən]*n.*（大）公司
investigation[inˌvesti'geiʃən]*n.*科学研究;学术研究
parlance['pɑ:ləns]*n.*说法;用语
colloquially[kə'ləukwi:əl]*adv.*口语化;口头化（地）
lingo['liŋgəu]*n.*术语;行话
trail[treil]*n.*（长串的）踪迹;痕迹
registrar['redʒiˌstrɑ:]*n.*（大学的）教务长,教务主任
profound[prə'faund]*adj.*巨大的;深远的
parse[pɑ:s]*v.*（对句子）作语法分析;作句法分析
extract[iks'trækt]*v.*提取;抽取
underlying[ˌʌndə'laiiŋ]*adj.*底层的
partition[pɑ:'tiʃən]*v.*分割;隔开
statistic[stə'tistik]*n.*统计数据
atomically[ə'tɔmikəli]*adv.*整体地;不可再分地
isolation[ˌaisə'leiʃən]*n.*隔离;隔离状态
guarantee[ˌgærən'ti:]*n.*保障
algebra['ældʒibrə]*n.*代数

L9

dimensional（构成形容词）…维的
illustrate['iləstreit]*v.*（用示例、图表等）说明;解释
parenthesized[pə'renθisaizd]*adj.*补充的;插入说明的
comma['kɔmə]*n.*逗号
surround[sə'raund]*v.*环绕;围绕
colon['kəulən]*n.*冒号

续表

immaterial[ˌiməˈtiəri:əl]*adj.*无关紧要的;不相干的	
permute[pə(:)ˈmju:t]*v.*改变序列,组合;置换	
static[ˈstætik]*adj.*静态的;不变化的	
existing[igˈzistiŋ]*adj.*现存的;存在的	
expel[iksˈpel]*v.*驱逐;赶走	
declare[diˈklɛə]*v.*表明;宣传;断言	
indicate[ˈindikeit]*v.*标示;显示（信息）	
imagine[iˈmædʒin]*v.*想象;设想	
assert[ɔˈsɔ:t]*v.*明确肯定;断言	

L10

channel[ˈtʃænl]*n.*途径;渠道;系统	
coordinate[kəuˈɔ:dineit]*v.*协调;使…相配合	
sharing[ˈʃɛəriŋ]*n.*共享	
network[ˈnetwə:k]*v.*（计算机）将…连接成网络	
integrate[ˈintigreit]*v.*使合并,成为一体	
self-contained[ˌselfkənˈteind]*adj.*自治的	
dispersed[disˈpə:st]*adj.*分治的	
geographical[dʒiəˈgræfikəl]*adj.*地理上的	
peripheral[pəˈrifərəl]*adj.*外围的;非核心的	
handle[ˈhændl]*v.*处理;应对	
destination[ˌdestiˈneiʃən]*n.*目的地	
autonomously[ɔ:ˈtɔnəməs]*adv.*自治地;自主地	
occasionally[əˈkeiʒənəli]*adv.*偶尔,间或	
itinerary[aiˈtinəˌreri:]*n.*行程;旅行日程	
evolve[iˈvɔlv]*v.*逐步发展;逐步演变	

L11

internetwork[ˌintəˈnetˌwə:k]*v.*网络互联	
individual[ˌindiˈvidjuəl]*n.*个人	
monthly[ˈmʌnθli]*adj.*按月计算的	
fee[fi:]*n.*费用	
permanent[ˈpə:mənənt]*adj.*持续的;固定的	

purposeful['pə:pəsfəl]*adj.*有目的的;有意图的;有意义的	
vulnerable['vʌlnərəbl]*adj.*脆弱的;易受影响的	
offset['ɔfset]*v.*补偿;抵消;耗费	
route[ru:t]*v.*（按一定的线路）传递;发送	
dedicate['dedikeit]*v.*把…奉献给	
retrieve[ri'tri:v]*v.*检索（存储的信息）	
offer['ɔfə]*v.*提供;供应	
productivity[ˌprɔdʌk'tiviti]*n.*生产率;生产效率	
facilitate[fə'siliteit]*v.*促进;使便利	
facility[fə'siliti]*n.*特色服务	
participant[pɑ'tisipənt]*n.*参与者;参加者	
portion['pɔ:ʃən]*n.*（一）部分	
indicator['indiˌkeitə]*n.*标识	
radiology[ˌreidi:'ɔlədʒi:]*n.*放射学;放射医疗	
conceivable[kən'si:vəbl]*adj.*能想到的;可想象的	
subscriber[səb'skraibə]*n.*消费者;用户	
simultaneously[saiməl'teiniəsli]*adv.*可同时进行地;同步地	
purchase['pə:tʃəs]*n.*购买;消费	
corporation[ˌkɔ:pə'reiʃən]*n.*（大）公司	
nonprofit[nɔn'prɔfit]*adj.*非营利的	

L12

vandalism['vændlˌizəm]*n.*恣意破坏;故意破坏	
attack[ə'tæk]*v.*攻击;袭击	
incorporate[in'kɔ:pəreit]*v.*包含;包括	
malicious[mə'liʃəs]*adj.*恶意的;恶毒的	
infect[in'fekt]*v.*传染,使感染（计算机病毒）	
reside[ri'zaid]*v.*居住在;定居于	
devastating['devəsteitiŋ]*adj.*毁灭性的	
corrupt[kə'rʌpt]*v.*引起（计算机文件等的）错误;破坏	
autonomous[ɔ:'tɔnəməs]*adj.*自治的	
forward['fɔ:wəd]*v.*发送;转寄	
replicated['repliˌkeit]*adj.*再生的;自我复制的	
legitimate[li'dʒitimit]*adj.*正当的;合理的	

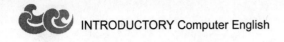

disguise[disˈgaiz]v.伪装;假扮	
victim[ˈviktim]n.受害者;受骗者	
dormant[ˈdɔ:mənt]adj.潜伏的;蛰伏的	
trigger[ˈtrigə]v.触发;引起	
enticing[inˈtaisiŋ]adj.有诱惑力的;诱人的	
misdeed[misˈdi:d]n.恶行;不义之举	
instigator[ˈinstigeitə]n.（幕后）唆使者	
blatantly[ˈbleitəntli]adv.（坏的行为）明目张胆地;公然地	
explicitly[iksˈplisitli]adv.不隐晦地;直截了当地	
con[kɔn]n.骗局;诡计	
perpetrator[ˌpə:piˈtreitə]n.作恶者;犯罪者	
hostile[ˈhɔstail]adj.恶意的;敌对的	
launch[ˈlɔ:ntʃ]v.发动;发起（尤指有组织的活动）	
disrupt[disˈrʌpt]v.扰乱;中断	
halt[hɔ:lt]n.暂停;停止	
swamp[swɔmp]v.使不堪承受;疲于应对	
accomplice[əˈkɔmplis]n.从犯;帮凶	
intruder[inˈtru:də]n.侵入者;闯入者	
proliferation[prəuˌlifəˈreiʃən]n.（迅速）繁殖;猛增	
junk[dʒʌŋk]n.垃圾	
overwhelm[ˌəuvəˈhwelm]v.压垮;使应接不暇	
compound[ˈkɔmpaund]v.使加重;使恶化	
detrimental[ˌtetriˈmentl]adj.有害的;不利的	
adage[ˈædidʒ]n.谚语;格言	
terminate[ˈtə:mineit]v.（使）结束;终止	
masquerade[ˌmæskəˈreid]v.伪装;假扮	
clandestine[klænˈdestin]adj.暗中的;秘密的	
preventative/preventive[priˈventətiv]adj.预防性的;防备的	
connotation[ˌkɔnəˈteiʃən]n.含义;隐含意义	
shield[ʃi:ld]v.保护;庇护	
detect[diˈtekt]v.发现;发觉	
irregularity[iˌregjəˈlæriti:]n.不正当的行为;不正常的做法	
invasion[inˈveiʒən]n.侵入;侵犯	
vaccine[vækˈsi:n]n.疫苗	
routinely[ru:ˈti:nli]adv.常规地;日常地	

L13

interrelated[ˌintəriˈleitid]*adj.*相互关联的
retrieve[riˈtri:v]*v.*（计算机的）检索数据
visualize[ˈviʒuəlaiz]*v.*使形象化
capture[ˈkæptʃə]*v.*采集;收集
convert[kənˈvə:t]*v.*使转变,转换
appropriate[əˈprupriət]*adj.*相应的;恰当的
evaluate[iˈvæljueit]*v.*评价;评估
correct[kəˈrekt]*v.*纠正;修正
massive[ˈmæsiv]*adj.*大量的
repository[riˈpɔziˌtɔ:ri:]*n.*（正式）大仓库;大储物处
fixed[fikst]*adj.*固定的;不变的
disseminate[diˈsemineit]*v.*传播,散布（知识、信息等）
conformity[kənˈfɔ:miti]*n.*遵从,遵守
identify[aiˈdentifai]*v.*标识;显示;说明
manual[ˈmænjuəl]*adj.*手工的;手动的
sharp[ʃɑ:p]*adj.*清晰的;鲜明的
analogy[əˈnælədʒi]*n.*类比;比拟
hammer[ˈhæmə]*n.*锤子

L14

conduct[kənˈdʌkt]*v.*组织;安排
confront[kənˈfrʌnt]*v.*使…无法回避;降临于…
underlying[ˌʌndəˈlaiiŋ]*adj.*根本的;基础的
vast[vɑ:st]*adj.*巨大的;广阔的
exchange[iksˈtʃeindʒ]*v.*交换;互换
furiously[ˈfjuəriəsli]*adv.*猛烈的;激烈的;疯狂的
brochure[ˈbrəuʃuə]*n.*资料（或广告）手册
manual[ˈmænjuəl]*n.*使用手册;说明书
mall[mɔ:l]*n.*购物广场;大卖场
stock[stɔk]*n.*股票
bond[bɔnd]*n.*债券
mutual[ˈmju:tʃuəl]*adj.*共有的;共同的

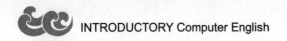
续表

| fuel[fjuəl]v.加强;刺激 |
| encompass[en'kʌmpəs]v.包含;涉及（大量事物） |
| alternative[ɔ:l'tə:nətiv]n.替代物;替换品 |
| accelerate[æk'seləreit]v.使加速;加快 |
| inventory['invəntri]n.（商店的）库存;存货 |
| facilitate[fə'siliteit]v.促进;促使;使便利 |
| soar[sɔ:]v.猛增;急升 |
| far-flung['fɑ:'flʌŋ]adj.分布广的;广泛的 |
| seamlessly['si:mlisli]adv.无缝地 |
| redefine[,ri:di'fain]v.重新定义 |
| reinvent[,ri:in'vent]v.以新形式出现 |

L15

| encompass[en'kʌmpəs]v.包含;涉及（大量事物） |
| discrete[dis'kri:t]adj.分离的;互不相连的 |
| sketch[stetʃ]n.概述;简述 |
| significant[sig'nifikənt]adj.重要的;有显著意义的 |
| rendering['rendəriŋ]n.渲染 |
| depict[di'pikt]v.描写;描述 |
| sophisticated[sə'fistikeitid]adj.高级的;先进的 |
| model[mɔdəl]v.建模 |
| realistic[,riə'listik]adj.真实的;逼真的 |
| obey[əu'bei]v.遵守;遵循 |
| narrative['nærətiv]n.描述;叙述 |
| constrain[kən'strein]v.限制;约束 |
| suspension[sə'spenʃən]n.延迟;延缓 |
| gravity['græviti]n.重力;地心引力 |
| scoring['skɔ:riŋ]n.计算得分 |
| combination[,kɔmbi'neiʃən]n.联合体;综合体 |
| overlay[,əuvə'lei]v.在…上;覆盖 |
| soul[səul]n.灵魂 |
| tromp/tramp[trɔmp]v.（尤指长时间地）重步行走;踏;踩 |
| terrain[te'rein]n.地形 |
| lamppost['læmp,pəust]n.顶杆;路灯柱 |

续表

imaginative[iˈmædʒinətiv]*adj.*富于想象的
flavoring[ˈfleivəriŋ]*n.*调味品;调味香料
cue[kju:]*n.*（戏剧的）暗示;提示;尾白
judicious[dʒuːˈdiʃəs]*adj.*审慎而明智的;有见地的
ponder[ˈpɔndə]*v.*考虑;琢磨
patch[pætʃ]*n.*补丁文件
accompany[əˈkʌmpəni]*v.*伴随
launch[lɔ:ntʃ]*v.*发动;发起（尤指有组织的活动）
vibrant[ˈvaibrənt]*adj.*充满生机的;生气勃勃的
community[kəˈmjuːniti]*n.*社团方式
fiddle[ˈfidl]*v.*篡改;伪造
authenticate[ɔːˈθentiˌkeit]*v.*证明…是真实的、有效的;证实

L16

rig[rig]*v.*绑定
realism[ˈriːəˌlizəm]*n.*真实性;逼真
intimately[ˈintimitli]*adv.*密切地;紧密地
blink[ˈbliŋk]*v.*眨眼
mallet[ˈmælit]*n.*木槌
muscle[ˈmʌsl]*n.*肌肉
specific[spiˈsifik]*adj.*特定的
appropriately[əˈprəupriˌeitli]*adv.*相应地;恰当地
adjust[əˈdʒʌst]*v.*调整;调节
diagnose[ˈdaiəgnəuz]*v.*判断（问题的原因）
plot[plɔt]*v.*（在坐标图上）画出,标出
spot[spɔt]*v.*看出;注意到;发现
tweak[twiːk]*v.*稍稍调整
interruption[ˌintəˈrʌpʃən]*n.*中断
glitch[glitʃ]*n.*小差错;小故障
bump[bʌmp]*n.*碰撞;撞击
glue[gluː]*n.*胶水
commercial[kəˈmɔːʃəl]*n.*（电台或电视播放的）广告
feature[ˈfiːtʃə]*n.*（电影的）正片,故事片
verse[vəːs]*n.*歌曲的段落

chorus['kɔ:rəs]*n.*副歌	
perceive[pə'si:v]*v.*感觉到;察觉到;意识到	
sequentially[si'kwenʃəli]*adv.*按次序地;从而	
recoil[ri'kɔil]*v.*退缩;畏缩	
simultaneously[saiməl'teiniəsli]*adv.*同时进行地;同步地	
exception[ik'sepʃən]*n.*例外的事物;规则的例外	
kinematics[kini'mætiks]*n.*运动学	
anatomically[ˌænə'tɔmikəli]*adv.*从剖析的角度;从解析的角度	
limb[lim]*n.*肢;臂;腿	
flexible['fleksəbl]*adj.*柔韧的;可弯曲的;有弹性的	
gymnast['dʒimˌnæst]*n.*体操运动员	
absorb[əb'sɔ:b]*v.*减轻（打击、碰击等的）作用	
momentum[məu'mentəm]*n.*冲力	
lean[li:n]*v.*前俯	
exaggerate[ig'zædʒəreit]*v.*夸张;夸大	
crisp[krisp]*adj.*清晰分明的;简明扼要的	
perfect['pə:fikt]*adj.*正合适的	
relaxed[ri'lækst]*adj.*（人）放松的;自在的	

Key Terms Index

L1

application software	memory
arithmetic and logic unit	memory unit
central processing unit	operating system
computer	output device
computer program	output unit
computer programmer	personal computer
data	primary memory
disk	random access memory(RAM)
hard drive	secondary storage unit
hardware	software
input device	supercomputer
input unit	system software
logical unit	

25

L2

alignment	information
American National Standards Institute	line
American Standard Code for Information	line feed
amplitude	lowercase letter
audio	Musical Instrument Digital Interface
binary notation	numerical data
bit	pixel
bit map	pixel's luminance
bit pattern	RGB encoding
black and white image	sample rate
blue chrominance	sound
carriage return	sound wave

color image	tab
computer-aided design	text
control information	text editor/simply editor
curve	text file
digit	three-dimensional object
dot	two's complement notation
file	Unicode
floating-point notation	uppercase letter
font	vector
image	word processor

44

L3

applications program	kernel
client operating system	memory space
computer architecture	multiuser system
computer system	network operating system(NOS)
control program	network server
CPU time	operating system
desktop operating system	resource allocator
embedded operating system	spell checker
file storage space	stand-alone operating system
hardware	user
I/O device	

21

L4

software system	processing function
systems development	business procedure
systems development process	data model
requirement change	control
system analysis	physical design

续表

information requirement	programming
system solution	testing
analyst	unit testing
feasibility study	system testing
technical feasibility	acceptance testing
economic feasibility	test plan
operational feasibility	conversion
requirements analysis	parallel strategy
systems failure	direct cutover strategy
systems development cost	pilot study strategy
system design	phased approach strategy
system designer	documentation
logical design	production
input	technical specialist
output	maintenance

40

L5

coding phase	lookup table
data structure	associative array
array	dictionary
element	index type
subscript operator []	index variable
stack	key
last-in-first-out(LIFO)	value
queue	graph
first-in-first-out(FIFO)	vertex
list	node
random access	edge
tree	arc
nonlinear	terminal point
successor	endpoint
children	Abstract Data Type(ADT)
root	linked list

续表

parent	primitive type
predecessor	data item
table	permissible operation
map	

39

L6

flowchart	class
top-down design	attribute
procedure	behavior
main procedure	field
procedural programming	method
procedural programming language	reference
parameter	software building block
fundamental programming	reusable
object-oriented programming(OOP)	chip
object	

19

L7

high-level programming language	dynamic
Oak	pointer
embedded consumer electronic appliance	multiple inheritance
Java	interface
Internet application	automatic memory allocation
Web programming	garbage collection
standalone application	procedural language
server	Object-oriented programming(OOP)
desktop	procedural programming
mobile device	procedure
distributed application	object
the Internet	encapsulation
Web browser	inheritance

applet	polymorphism
graphical user interface	distributed computing
Java servlet	interpreter
JavaServer Page(JSP)	bytecode
dynamic Web page	Java Virtual Machine(JVM)
hand-held device	compiler
Java-language white paper	machine code
simple	native machine
object-oriented	reliable
distributed	runtime exception-handling
interpreted	exception
robust	runtime error
secure	security mechanism
architecture-neutral	platform-independent
portable	multithreading
multithreaded	software version

58

L8

database	secondary storage
database management system(DBMS)	main memory
database system	storage manager
schema	buffer manager
query	buffer
modify	page-sized region
query language	disk block
durability	data
isolation	log record
atomicity	statistic
system component	index
data structure	transaction
control	transaction manager
data flow	transaction command
database administrator(DBA)	logging

table	concurrency control
relation	deadlock resolution
column	query processor
schema-altering command	query compiler
data-definition language(DDL)	query plan
DDL processor	query parser
execution engine	query preprocessor
index/file/record manager	semantic check
metadata	query optimizer
data-manipulation language(DML)	scheduler

50

L9

relational model	row
table	tuple
relation	integer
column	string
attribute	domain
entry	data type
schema	column header
set	instance
list	current instance
database	key
relational database schema	artificial key
database schema	

23

L10

network	mainframe
computer network	network's topology
network architecture	star network
node	polling

client	time-sharing system
microcomputer	bus network
server	bus
file server	ring network
printer server	distributed data processing system
communication server	hierarchical network
Web server	hybrid network
database server	strategy
network operating system(NOS)	terminal network system
distributed processing	terminal
decentralized organization	peer-to-peer network system
main computer	authority
centralized computer	client/server network system
host computer	enterprise computing
minicomputer	

37

L11

the Internet	e-mail address
the Net	@ symbol
modem	name
Internet Service Provider(ISP)	identifier
online service	domain name
TCP/IP(Transmission Control Protocol/ Internet Protocol)	subdomain
client/server technology	top level domain
individual	com
graphical user interface	gov
character-character product	country indicator
e-mail	edu
Usenet newsgroup	electronic bulletin board
LISTSERV	e-mail mailing list server
chatting	Internet Relay Chat(IRC)
Telnet	chat group

FTP	channel
gopher	information retrieve
the World Wide Web	file transfer protocol(FTP)

36

L12

malware	firewall
virus	spoofing
worm	spam filter
Trojan horse	proxy server
spyware	auditing software
sniffing	utility package
phishing	antivirus software
denial of service attack	update
spam	email attachment
filter	pop-up add

20

L13

information system	output
decision making	processed information
control	feedback
coordination	computer-based information system(CBIS)
information	formal system
data	office gossip network
input	manual system
raw data	operating instruction
processing	designing solution
meaningful form	

19

L14

information system	electronic commerce
the Internet	information flow
electronic market	Internet technology
information	personnel policy
product	account balance
service	production plan
payment	require
electronic middlemen	maintenance
cost	design document
marketplace transaction	intranet
purchase	electronic business
sale	manager
global marketplace	e-mail
the World Wide Web	Web document
the Web	work-group software
retailer	supplier
electronic shopping mall	business partner
customer	business model
manufacturer	business process
product type	corporate culture
financial trading	relationship
business-business transaction	

43

L15

3D game	mouse cursor
game engine	Torque Game Engine
gaming environment	model
scene rendering	terrain
networking	texture
graphic	sound
scripting	sound effect

续表

game environment	multiplayer game
textured polygon rendering	music
First-Person Shooter(FPS)	single-player adventure game
graphic environment	story line mood
character	contextual cue
physics formula	game infrastructure
natural physical law	Web site
script	auto-update program
play function	log in
Graphical User Interface(GUI)	community forum
control input	bulletin board
Heads Up Display(HUD)	feedback mechanism
upper-left corner	administrative tool
GUI text control	persistence
GUI button control	database back end
logo	database
GUI bitmap control	

47

L16

animation	dope sheet
animator	language of movement
physics	vocabulary of motion
motion	arc
time	multiple force
raw material	mechanical motion
animation timing	joint
timing	forward kinematics
illustration	inverse kinematics
mood	straight line
personality	motion curve
force	drag
hand-drawn animation	multijointed object
assistant animator	compress

key	squash
keyframe	stretch
inbetween	shape
motion graph	anticipation
horizontal axis	zip out
vertical axis	overshoot

40

Bibliography

[Brookshear]

Computer Science: An Overview,9E, by J. Glenn Brookshear.

Addison-Wesley Publishing Company, Reading, MA(2006)0-32-138701-5.

[Bousquet]

Model, Rig, Animate with 3ds Max 7,1E, by Michele Bousquet.

New Rider Press, Berkeley, CA(2005)0-32-132178-2.

[Clinton]

Game Character Modeling and Animation with 3ds Max,1E, by Yancey Clinton.

Focal Press, Burlington, MA(2007)0-24-080978-5.

[Comer]

Computer Networks and Internets,5E, by Douqlas E. Comer.

Prentice Hall, Upper Saddle River, NJ(2008)0-13-606698-4.

[Date]

An Introduction to Database Systems,8E, by C. J. Date.

Addison-Wesley Publishing Company, Reading, MA(2003)0-32-119784-4.

[Feil]

Beginning Game Level Design,1E, by John Feil and Marc Scattergood.

Course Technology PTR, Boston, MA(2005)1-59-200434-2.

[Finney]

3D Game Programming All in One,2E, by Kenneth C. Finney.

Course Technology PTR, Boston, MA(2006)1-59-863266-3.

[Haines]

Java^{TM} 2 Primer Plus,1E, by Steven Haines, Stephen Potts.

Sams Publishing, Indianapolis, IN(2002)0-67-232415-6.

[Halsall]

Computer Networking and the Internet,5E, by Fred Halsall.

Addison-Wesley Publishing Company, Reading, MA(2005)0-32-126358-8.

[Hewings]

Advanced Grammar in Use,2E, by Martin Hewings.

Cambridge University Press, New York, NY(2005)0-52-161403-1.

[Hubbard]

Programming with C++,2E, by John R. Hubbard.

McGraw-Hill, New York, NY(2000)0-07-135346-1.

[Laudon]

Management Information Systems: Managing the Digital Firm,9E, by Kenneth C. Laudon and Jane P. Laudon.

Prentice Hall, Upper Saddle River, NJ(2005)0-13-153841-1.

[Lafore]

Data Structures and Algorithms in Java,2E, by Robert Lafore.

Sams Publishing, Indianapolis, IN(2002)0-67-232453-9.

[Levitin]

Introduction to the Design and Analysis of Algorithms,2E, by Anany V. Levitin.

Addison-Wesley Publishing Company, Reading, MA(2006)0-32-135828-8.

[Liang]

Introduction to Java Programming,Comprehensive,8E, by Y. Daniel Liang.

Prentice Hall, Upper Saddle River, NJ(2010)0-13-213080-7.

[Luebke]

Level of Detail for 3D Graphics,1E, by David Luebke, Martin Reddy, Jonathan D. Cohen, Amitabh Varshney, Benjamin Watson, and Robert Huebner.

Morgan Kaufmann, San Francisco, CA(2002)1-55-860838-9.

[Maestri]

Digital Character Animation 3(No.3),1E, by George Maestri.

New Rider Press, Berkeley, CA(2006)0-32-127600-5.

[McCarthy]

English Vocabulary in Use Advanced,1E, by Michael McCarthy and Felicity O'Dell.

Cambridge University Press, New York, NY(2006)0-52-167746-7.

[Murdock]

3ds Max 2009 Bible,1E, by Kelly L. Murdock.

Wiley Publishing, Inc., Indianapolis, IN(2009)0-47-047191-3.

[Nutt]

Operating Systems,3E, by Gary Nutt.

Addison-Wesley Publishing Company, Reading, MA(2004)0-21-177344-9.

[O'Dell]

English Idioms in Use Advanced,1E, by Felicity O'Dell and Michael McCarthy.

Cambridge University Press, Cambridge, CB2(2010)0-52-174429-6.

[O'Leary]

Computing Essentials 2010 Introductory Edition,20E, by Timothy J. O'Leary and Linda I. O'Leary.

John Wiley & Sons, New York, NY(2009)0-07-351674-0.

[Panko]

Corporate Computer and Network Security,2E, by Raymond Panko.

Prentice Hall, Upper Saddle River, NJ(2009)0-13-185475-5.

[Penton]

Data Structures for Game Programmers,1E, by Ron Penton.

Premier Press, Cincinnati, Ohio(2002)1-93-184194-2.

[Raposa]

Java in 60 Minutes A Day,1E, by R. F. Raposa.

John Wiley & Sons, New York, NY(2003)0-47-142314-9.

[Scott]

Programming Language Pragmatics,3E, by Michael L. Scott.

Morgan Kaufmann, San Francisco, CA(2009)0-12-374514-4.

[Sharp]

Microsoft Visual J# .NET,1E, by John Sharp, Andy Longshaw, and Peter Roxburgh.

Microsoft Press, Redmond, WA(2002)0-73-561550-0.

[Silberschatz]

Operating System Concepts,8E, by Abraham Silberschatz, Peter B. Galvin, and Greg Gagne.

McGraw-Hill, New York, NY(2008)0-47-012872-0.

[Stallings]

Computer Organization and Architecture:Designing for Performance,8E, by William Stallings.

Prentice-Hall, Upper Saddle River, NJ(2009)0-13-607373-5.

[Tanenbaum]

Modern Operating System,3E, by Andrew S. Tanenbaum.

Prentice-Hall, Upper Saddle River, NJ(2007)0-13-600663-9.

[Trumble]

Shorter Oxford English Dictionary,6E, by William R. Trumble.

Oxford University Press, New York, NY(2007)0-19-923324-1.

[Ullman]

A First Course in Database Systems,3E, by Jeffrey D. Ullman and Jennifer Widom.

Prentice-Hall, Upper Saddle River, NJ(2007)0-13-600637-X.

[Weiss]

Data Structures and Algorithm Analysis in C++,3E, by Mark A. Weiss.

Addison-Wesley Publishing Company, Reading, MA(2006)0-32-144146-X.